Introduction to Quality
Control, Assurance, and Management

GREGORY B. HUTCHINS

**Merrill, an imprint of
Macmillan Publishing Company**
New York

Collier Macmillan Canada, Inc.
Toronto

Maxwell Macmillan International Publishing Group
New York   Oxford   Singapore   Sydney

I sincerely thank George and Irina Hutchins and Baba.

Executive Editor: Stephen Helba
Production Editor: Sharon Rudd
Art Coordinator: Vincent A. Smith
Cover Designer: Brian Deep

This book was set in Times Roman.

Macmillan Publishing Company
866 Third Avenue, New York, NY 10022

Collier Macmillan Canada, Inc.

Library of Congress Catalog Card Number: 90-53390
International Standard Book Number: 0–675–20896–3

Printing: 1 2 3 4 5 6 7 8 9   Year: 1 2 3 4

Preface

The challenges, excitement, and opportunities in quality technology and management are increasing yearly. In the 1990s, all organizations face the issues of improving product and service quality, containing costs, and enhancing innovation. Quality and competitiveness are now synonymous.

The discipline of quality has evolved and expanded rapidly from inspection to company-wide quality management. Not too many years ago, quality was inspected into a product after it was made. Then quality was controlled in the manufacturing process. Now quality is a company-wide phenomenon encompassing every person and every element of the organization.

Because quality topics and technologies have increased as quickly as the importance of the field, this book covers traditional topics in inspection, quality control, and quality assurance as well as current topics in company-wide quality management. This text covers the mechanics of quality problem solving while emphasizing the importance of quality management and quality decision making.

Throughout the book, I have tried to present important technical material in a fun, readable, simple, stimulating, and instructive manner. I have tried to enliven a very important topic and show its relevance in a number of areas, as well as to cover the fundamental technologies. I believe that if technical concepts are made relevant and interesting, they can be learned more easily and remembered longer.

Introduction to Quality: Management, Assurance, and Control is designed for a first, one quarter course in quality for students majoring in technology and engineering. However, the coverage and level of topics may make it appropriate for students as an introduction to any quality course in any field.

My primary objective is to introduce, integrate, and explain quality management, assurance, and control principles in the clearest possible manner, and to provide students with an introductory text that will assist them in applying quality analysis to real-world problems.

The mathematical prerequisite is a basic algebra course. Computational procedures are presented and demonstrated only to the extent that they help students understand quality principles. Formulas are derived intuitively and are presented in their simplest form.

Within each chapter, difficult quality subjects begin with a discussion and are followed by an industry example and finally by a numerical example. Progressively more difficult topics are gradually introduced.

Special features of the text are:

- ☐ **Real-life examples.** The importance and application of statistical techniques are illustrated by real-life examples.
- ☐ **Key concept emphasis.** Key concepts of customer satisfaction, management of quality, defect prevention, and continuous improvement are constantly emphasized. They are incorporated into the text and are highlighted in the Spotlight special features.
- ☐ **Key word highlights.** Key words are boldfaced throughout the text, explained in context, and defined succinctly in the Key Terms section at the end of each chapter.
- ☐ **Concept summary.** Key concepts are reviewed and summarized at the end of each chapter.
- ☐ **Discussion questions.** At the end of each chapter, problems emphasize computational skills, while discussion questions stress conceptual understanding. These sections can serve as the basis for classroom discussions or for homework assignments.

I thank Steve Helba and Sharon Rudd for their unwavering support of this project.

My sincerest thanks go to the following for their insightful and tough reviews and comments: Miles Weaver, Terra Technical College; Clarence Fauber, Indiana State University; Harold Hambrock, DeVry–Columbus; John Vittrup, Southwest Texas State University; Robert Homolka, Kansas College of Technology; Kurt Blumberg, Milwaukee Area Technical College; John Troche, University of Akron; Joseph Thompson, North Central Technical College; David Lyth, Western Michigan University; James Wertz, Aiken Technical College; Michael Bowman, Purdue University of Indianapolis; Tom Lavender, Catawba Valley Technical Institute; Timothy Sexton, Ohio University–Athens; Larry Roderick, Texas A & M University; Gary Winek, Southwest Texas State University; Steve Redmer, Lakeshore Technical College; and Saeid Eidgahy, Bowling Green State University.

You, the instructor and students, are my customers. I want to hear from you. Please send your comments to me or to the publisher.

Greg Hutchins
Quality Plus Engineering
Portland, Oregon

About the Author

Greg Hutchins was an instructor at Portland Community College in Mathematics and Technology. For the past five years he has been a principal with Quality Plus Engineering in Portland, Oregon. Quality Plus Engineering is a quality engineering and management firm that provides consulting services to Fortune 500 companies and small businesses.

Contents

6
CONTROL CHARTS   130

APPENDICES

INDEX   287

1

Quality Management

In the global marketplace, the issue of quality is changing, continuously adapting to customer needs and expectations. Several important trends are accelerating this change.

Customer demand for quality is increasing. Quality is being enhanced through increased product performance, design, usability, reliability, and maintainability.

Defect prevention is replacing defect inspection as the means to pursue continuous quality improvement. Many organizations are eliminating incoming, in-process, and final inspection. Responsibility for quality is being placed on the person doing the work, whether assembling, fabricating, managing, servicing, or delivering.

Company-wide quality management is replacing quality assurance. Quality management is a broader concept than quality assurance or control. It not only implies controlling and assuring, but it also includes organizing, monitoring, coordinating, and even "cheerleading" quality.

WHAT IS QUALITY?

The term **quality** can be defined in various ways, depending on the perspective of the user. Quality is

☐ Conformance to applicable specifications and standards

1

☐ Fitness for use
☐ Satisfaction of customer wants, needs, and expectations at a competitive cost

Conformance. Every organization, whether profit, nonprofit, manufacturing, service, private, or public, has **specifications** and **standards.** Organizations develop these to measure performance and to correct deviations from expected levels of performance. For example, in a manufacturing operation, specifications detail dimensional limits or physical attributes of a quality characteristic of a part. In a service operation, standards dictate approved methods of behavior or service.

Fitness for Use. Joseph Juran, an eminent authority on quality management, coined the phrase "fitness for use" to define quality. This is a market- or customer-based definition. A product or service is fit for use if it satisfies customer needs and requirements.

An interesting point is that a product might be fit for use in terms of satisfying the customer, but not conform to the specification. A surface finish specification was developed for a consumer product. The condition of the surface finish is important because it enhances the product's appearance and hence its marketability. The specification was written to include all surfaces, both external and internal. However, if the inside product surface is blemished, but it cannot be seen by a customer and does not adversely influence the buy decision, the nonconformance is accepted. So a product with a blemish may be fit for use if the blemish does not affect performance, safety, or marketability.

Customer Satisfaction at a Competitive Price. Another definition says that product or service quality is the producer's ability to satisfy customer needs while still being able to realize a profit. This definition has both a customer and a manufacturer orientation. While the customer is the reason for the organization's existence, the product manufacturer and service provider must still make a profit.

This definition focuses on satisfying the customer at a competitive price. Many customers will not purchase a product or service unless it is reasonably priced.

HISTORY OF QUALITY

Early Quality

Quality techniques were first used in ancient times. Four thousand years ago, the Egyptians measured the rocks used in their pyramids. Then the Greeks and Romans measured buildings and aqueducts to ensure they conformed to requirements. Later, craft guilds in Renaissance Europe specified, measured, controlled, and assured the quality of paintings, cloth, tapestries, sculpture, and architecture. To assure uniformity, guild students went through exhaustive apprenticeship programs overseen by accomplished masters.

Modern Quality

The quality function in modern organizations has evolved through the following stages: inspection, quality control, quality assurance, and company-wide quality management.

Inspection. Modern quality started in the 1920s. The first quality groups were **inspection** departments. During production, inspectors measured products against specifications. Inspection departments were not independent; they usually reported to the manufacturing department whose efforts they were inspecting. This presented a conflict of interest. If the inspection department rejected a batch of nonconforming products and the manufacturing department wanted to push this batch of products out the door regardless of quality, the manufacturing department always got its way. This sent a "production at any cost" message to the organization instead of a "quality is job #1" message. Product quality could only improve slowly in this environment.

Quality Control. In the 1940s, inspection groups evolved into **quality control (QC)** departments. The start of World War II required that military products be defect-free. Product quality was crucial to winning the war and could only be ensured if the inspection department could control production processes. Quality, defined as conformance to specification, was controlled during production instead of being inspected into products. Responsibility for quality was transferred to an independent QC department, which was now considered the "guardian" of quality. Also, the QC department was now separated from manufacturing to give it autonomy and independence.

Quality Assurance. Quality control evolved into **quality assurance (QA).** The QA department focuses on assuring process and product quality through executing operational audits, supplying training, performing technical analysis, and advising operational areas on quality improvement. QA consults with the departments where the responsibility for quality actually rests.

QC is still alive in some organizations where QA has not evolved. It is considered to be a functional area, which is responsible for inspecting products, calibrating instruments, testing products, and inspecting incoming material.

Company-Wide Quality Management. As the issue of quality becomes more prominent, QA is evolving into a **company-wide quality management (CWQM)** function. CWQM is also called total quality management (TQM) or total quality control (TQC). The quality organization is the prime facilitator and consultant in this effort. Corporate quality groups are small with more authority but less direct responsibility for quality. For example, the quality organization has authority to stop defective material from leaving the manufacturing door, while the responsibility for the control of quality is pushed to the manufacturing department operator.

The chief executive officer (CEO) often starts and guides the CWQM program. As the quality message permeates the organization, more people become involved, and slowly a quality ethic and culture develop. The focus of the program is company-wide, customer-oriented, and competitively driven.

Quality no longer resides in one department. It is a company-wide issue essential to the organization's survival. To produce a quality product or deliver a quality service requires the attention and commitment of everyone in the organization. It is the responsibility of the person doing the work, whether it is the receptionist greeting people, the manager supervising employees, the operator fabricating material, or the person delivering flowers. Every element in the organization, from the executive committee that establishes policy to the receptionist at the front desk, contributes to or detracts from the quality effort. The executive committee defines a realistic policy; line management establishes doable objectives; engineers design attractive, reliable, and functional products; receptionists are courteous and prompt; and operators produce defect-free products.

Customer orientation is essential in CWQM programs because the customers' needs change and the organizations must adapt to changing needs. Adapting means designing aesthetic products, producing defect-free products, and delivering products on time, at a profit. Most importantly, an organization must design, produce, and deliver what the customer wants, not what the organization thinks the customer wants.

COMPANY-WIDE QUALITY MANAGEMENT

In a global economy, product manufacturing and service delivery know no boundaries. Corporate management might reside in Germany. An automobile might be designed in Italy. Parts might be made anywhere in the world. The automobile might be assembled in Mexico. It might be marketed and serviced in the United States.

The auto, from conception to manufacturing to delivery, has to embody quality. The **International Standards Organization (ISO)** developed standards (9000–9004) so that there could be a common language and understanding of important terms and concepts in quality.

The American standard "Quality Management and Quality System Elements—Guidelines" (ANSI/ASQC Q94-1987), issued by the **American National Standards Institute (ANSI),** evolved from the ISO standards. This standard specifies the principal elements of a CWQM system. ANSI standards are technically equivalent to ISO standards.

This book is primarily concerned with the producer of products rather than the deliverer of services. However, in almost every example, a product manufacturer is also a service deliverer. When a meal is ordered in a restaurant, the service component of the meal is as important as the product component, which is the meal itself. The waitress delivers food. The atmosphere is conducive to conversation, and the restaurant has special activities for entertaining children.

If the quality of complex goods and services is to be controlled and assured, a CWQM program is developed. The goal of the program is to measure, detect, reduce, eliminate, and prevent quality deficiencies. Deficiencies can be defective products, discourteous service, late deliveries, or nonserviceable automobiles.

Quality Loop

The **American Society for Quality Control (ASQC)** developed quality standards that are guidelines for implementing a quality management program. This series of standards (ANSI/ASQC Q90–94) embodies most elements of the different approaches to quality management.

Specifically, we discuss quality management in terms of ANSI/ASQC Standard Q94, "Quality Management and Quality System Elements—Guidelines." This standard identifies the CWQM elements that form a **quality loop** from market identification to disposal (Figure 1.1). The principal elements of the quality loop are outlined in the following paragraphs.

Marketing and Market Research. Marketing's responsibility is to identify a market, identify customer requirements, develop a product brief, and establish a feedback system.

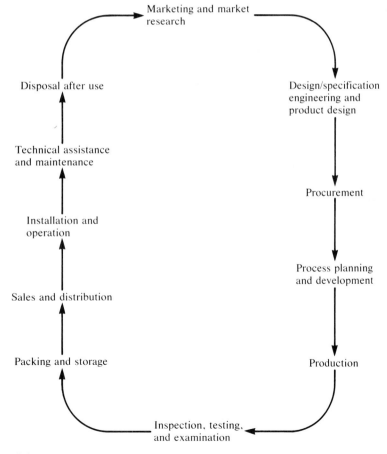

FIGURE 1.1
Quality loop. After American Society for Quality Control, ANSI/ASQC Standard Q94, "Quality Management and Quality System Elements—Guidelines."

Identify a market. Marketing takes the lead in identifying a market, its customers, and its requirements. A market can be an entirely new market, an established market, or a niche in an established market. A company might develop an entirely new product for a new market. This product is usually high priced in order to regain some of the research and development costs that were sunk into the project. This new product initially will not have any competition, so if customers want the product, they must pay its price.

For an established market, a company can produce an enhanced version that is sufficiently different from competing products, or it can develop a similar product and price it lower. This is a mass-marketing strategy based on lower priced products. Or a company can develop a specialized product for a niche in an established market. This product satisfies the needs of a specific target audience.

Identify customer requirements. Then marketing identifies customer needs, wants, and expectations in both products and services. Almost any product has a service component, which is as important as the product component. For example, people do not simply buy an IBM computer, they buy the service that is associated with the IBM name.

Develop a product brief. Once needs are identified, these are communicated to the organization in terms of a set of requirements distilled into a **product brief.** Customer requirements evolve into product and service specifications. A product brief can address all of the following: performance characteristics, aesthetics, serviceability, packaging, price, and regulatory requirements.

Establish a feedback system. Finally, marketing establishes an information, monitoring, and feedback system. Customer needs change, and an organization must continuously accommodate to changes throughout a **product's life cycle.** Otherwise products or services will age and not be able to satisfy current needs.

Most products follow a life cycle consisting of four stages: introduction, growth, maturity, and decline. These four stages show the sales pattern of a product. When a product is introduced in the marketplace, sales may be low because people are not aware of the product or do not know what benefits will accrue from its use, or the product may be priced too high. In its growth stage, through advertising people become aware of its potential benefits or of its ability to satisfy needs. Sales increase. In the mature stage, competitors develop enhanced products or similar lower priced products. Sales flatten out. Finally in the decline stage, competition forces the company to develop new products or lower the price of the existing product. Regardless, sales decrease. Figure 1.2 illustrates the product life cycle.

Design/Specification Engineering and Product Development. Engineering, using the product brief, translates customer requirements into technical specifications for materials, products, and processes. It considers the following in designing products: customer requirements, cost, manufacturability, testability, and design quality.

Customer requirements. Engineering first develops a concern or an idea and the knowledge of what the customer wants and expects. Often customer requirements, whether needs, wants, or expectations, are vague and engineering only has

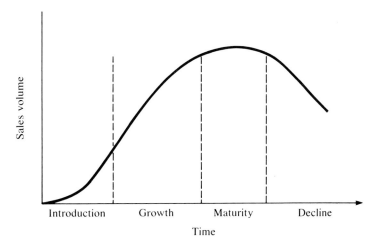

FIGURE 1.2
Product life cycle.

an abstract idea from marketing. Marketing obtains customer information from focus groups, surveys, polls, or other sources. Marketing believes that the organization can develop a product or service to satisfy these requirements. Engineering then determines whether it can develop the product within time and budget constraints.

Cost. While always considering customer satisfaction, product and delivery costs are never far from the designer's mind. If a product is high priced and has a quality image, it will sell to a narrow, affluent market, but usually not to a mass market.

Manufacturability. Sometimes engineers design products on a computer terminal and do not solicit input or advice with regard to **manufacturability** from manufacturing or purchasing. A product designed with tight tolerances is some-

SPOTLIGHT

Apple Corporation developed the first widely available commercial personal computer. It was expensive and demand was relatively small. Apple priced the computer high in order to recover the high start-up research and development costs.

Demand was initially low because the average person did not have an identified and daily use for the product. As software was developed, sales increased. Competition increased.

After several years, emulators cloned existing personal computers. These clones had similar performance characteristics but were lower priced. Competition on price, quality, and performance was fierce between such established companies as IBM, HP, Apple, Toshiba, NEC, and Samsung. So new companies entered the market, offering specialized or faster computers for specific market niches.

times inoperable or not manufacturable. Tight tolerances might result in mating parts interfering with each other, causing premature failure. A product designed with loose tolerances might result in unnecessary movement, again causing premature failure.

Testability. The ease and speed with which a product is accessible for testing is called **testability.** If the product is not positioned properly on a production line, it is difficult to test or even assemble. For example, in a cassette tape-recorder manufacturing facility, recorders travel on a conveyor belt where at set points they are assembled and tested. If the production operator has to physically move the product or perform any unnecessary movements, this represents wasted effort, which adds cost to the product.

Design quality. Quality in design is essential for a final product to be defect-free, safe, and reliable. Poor design can cause a product to fail prematurely. Also, if design errors are not eliminated, they are replicated in each manufactured product.

Good design also enhances safety and health. If a nuclear power plant is not designed and constructed with safety in mind, an accident can release radioactive waste in the atmosphere, resulting in loss of life. This was graphically illustrated in the Chernobyl and Three Mile Island accidents.

Design must also consider **reliability,** the long-term effects of quality. Asbestos was originally installed in buildings because it was considered a good insulator, but its dangers were not fully understood. Many years later it was discovered that asbestos is carcinogenic to those producing it, installing it, and living with it. The health risk was not known when engineers developed the product. Since then legal suits have resulted in massive financial liabilities to manufacturers, who many years ago were not aware of the risks.

Procurement. Depending on the product and industry, many manufacturers procure as much as 70% (in terms of the manufacturing dollar) of their material from outside suppliers. A final product is only as good as its procured components. Purchasing's responsibilities are to communicate requirements to suppliers, to select suppliers, and to monitor suppliers.

Communicate requirements to suppliers. A manufacturer can be thought of as being a customer to its suppliers. The customer must communicate requirements to suppliers explicitly through specifications, drawings, purchase orders, and contracts. At a minimum, requirements should include the following:

- ☐ Product quality characteristics
- ☐ Service characteristics
- ☐ Operating environment
- ☐ Precise identifications of aesthetics and quality grade
- ☐ Inspection instructions
- ☐ Product specifications

Select suppliers. Supplier selection is a formal process that evaluates suppliers based on their ability to produce a defect-free product, which is competitively priced, delivered on time, and serviced properly.

Monitor suppliers. Suppliers are monitored to ensure that they are providing products that conform to specifications. Monitoring techniques include on-site audits of the supplier's quality system, product testing, and product improvement projects.

Process Planning and Development. Production processes, whether inspection, assembly, or fabrication, should be planned so that they run smoothly under controlled conditions. This means that fabrication equipment, production equipment, and measuring instruments are monitored, calibrated, and controlled within prescribed operating limits so that defect-free products are produced. Dimensional dispersion caused by operators, materials, methods, and machines is kept to a minimum. Production operations are detailed. Work and maintenance instructions are followed. Line people are instructed in measurement.

Production. Production is responsible for creating reality out of engineering specifications and drawings. Production is a loose term that includes assembly, fabrication, and on-line inspection.

The responsibility for quality rests with the production supervisor and worker. The production supervisor must communicate the importance of quality to the person on the line. This person will only be committed if the organization is honestly dedicated to the pursuit of excellence and of never-ending improvement. If the organization's commitment is an expedient measure and not genuine, the line worker will sense manipulation, resent it, and may not be concerned about the quality she or he is producing.

Statistical process control. Production processes must be capable of manufacturing products. Once a process is in operation, control is maintained to ensure that the dimension of the quality characteristic remains inside process control action limits. Otherwise the operator or automatic measurement instruments will adjust the process so that it is again producing acceptable parts. This is called **statistical process control,** which is discussed in Chapter 4.

Inspection, Testing, and Examination. Inspection, **testing,** and examination are conducted on processes and products. The level of testing depends on the product, risk to the consumer, risk to the producer, cost, and regulations. For example, products involving public health or safety have associated risks to the producer and the consumer if the product should fail. This type of product may be governed by federal regulations which mandate extensive testing and inspection. The Food and Drug Administration (FDA) governs the testing, manufacturing, and distribution of pharmaceutical products. The Nuclear Regulatory Commission (NRC) dictates the construction, maintenance, operation, testing, and commissioning of nuclear power plants.

Product quality is only as accurate as the measurement instruments used to check a quality characteristic. This implies that any measurement instrument should be accurate and precise to provide management with sufficient confidence in the decisions and actions based on the measurements. Gages, instruments, and automatic test equipment therefore must be chosen to meet or exceed the requirements of the customer. They should be calibrated at regular intervals.

Product quality can be verified through incoming, in-process, and final inspection. Incoming material is examined as it comes through the door. Statistical process control is used to control in-process performance. Final inspection occurs just before the product is sent to inventory or to the customer.

The degree and the frequency of inspection depend on the importance of the quality characteristic and the capability of the process. If a product quality characteristic is essential to product performance, then the product may be 100% inspected.

Packing and Storage. Quality can only be maintained if products are packaged, stored, handled, and transported properly. Proper packaging protects contents against damage due to vibration, shock, heat, abrasion, and corrosion. Some products require special storage, handling, and packaging. Electronic components may have to be packaged in antistatic containers. Perishable foods are transported in refrigerated containers.

Sales and Distribution. Once a product is sold, it must be delivered to the customer intact and in an expedient and courteous manner. If a product is damaged in transit, the problem must be handled effectively and efficiently with the customer. Also, discourteous delivery people can create a harmful impression of the product even before it is used. It is important to monitor and correct this situation because it may adversely affect future sales.

Installation and Operation. Complex industrial products can require specialized tools, equipment, methods, or trained personnel to install and to operate. At a minimum, complex machinery and products have extensive manuals that describe safe installation and operation. Installation and operating documentation includes instructions for assembly, repair, installation, and operation, spare parts lists, and product servicing information.

Technical Assistance and Maintenance. Once a product is sold, the buyer of a technical or complex product might require technical assistance to maintain its operation. Technical assistance might be nothing more than answering questions on the phone. After-sales service for complex equipment might mean sending a trained person to repair, replace, or maintain a sophisticated piece of equipment. After-sales service creates product or brand loyalty, which enhances repeat product sales.

Disposal after Use. After use, a product has to be disposed of properly and safely. If the product no longer has a useful life, it is scrapped. However, a product can have its life extended if it is repaired. If it must be scrapped, this is done in such a way that it does not jeopardize safety, health, or environment. For example, federal environmental laws regulate the disposal of hazardous chemicals, such as PCBs.

After a product has gone through the quality loop, the loop will start again with another product, whether it is a new or an enhanced product. There are very few products or services that do not follow the loop because products must adapt to marketplace changes or die.

INSPECTION AND PREVENTION

Inspection Mode

The **prevention** of defects, nonconformance, defectives, and flaws is a basic philosophy of this book. Prevention lowers defects, improves processes, and eventually lowers costs. The differences between an inspection and a prevention operation are illustrated in Figure 1.3.

The inputs are the same to both the inspection and the prevention processes, namely: machine, operator, material, methods and environment. Each is a potential cause of dispersion or variation of dimensions. Fabrication or assembly operations process inputs by adding value to the product in each operational step.

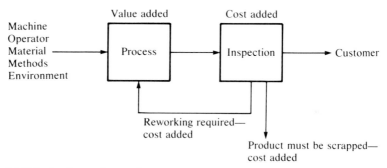

(a) Detection/inspection mode.

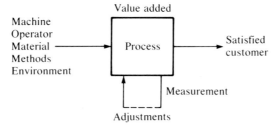

(b) Prevention/process control mode.

FIGURE 1.3
Comparison of inspection and prevention operations.

A process can be one step or 100 steps. At the end of the first step or the hundredth step, an operator might inspect or test a product to check its conformance with a specification. If acceptable, products are shipped to the customer. If rejected, products are scrapped or sent back through the process for rework. By the time a product has been fully processed, value has been added to the product in terms of direct labor, raw materials, equipment, and training. At this point if a product is scrapped or reworked, value has been lost [see Figure 1.3(a)].

Prevention Mode

In the prevention mode, the inputs are the same. Again, a process can be one step or 100 steps and value is added at each step. In the prevention mode, the process is controlled. The important difference between inspection and prevention modes is that during the process, the operator continually measures parts and adjusts the process if dimensions deviate from calculated limits [see Figure 1.3(b)].

Thus the operator is preventing defects form occurring by controlling the output of the operation. The objective is to send defect-free products to the customer.

MEASUREMENT

Quality Is Only as Good as the Measuring Instrument

Measurement is important because the quality of subsequent management decisions is only as reliable as the raw data obtained through the measurement instrument. Most quality analysis assumes that measurement data are accurate and precise. However, this may be an incorrect assumption because a measurement instrument may have been misused, abused, or not calibrated.

Just as the product of a manufacturing process can vary over time, the output of a complex measuring instrument can vary over time. Quality control is responsible for selecting the appropriate measuring instrument and maintaining its accuracy. Accuracy is established and maintained in a systematic program of periodic calibration, safe storage, and proper handling.

Police radar guns measure the speed of automobiles. Results from the radar gun are used as evidence for issuing speeding tickets. Most states have regulations that specify the calibration, handling, and storage of these sensitive instruments. Each day a radar gun has to be calibrated so that it provides accurate and repeatable measurements.

Accurate measurement requires that:

- ☐ Standard measurement units be used
- ☐ Instruments of required accuracy be selected to measure the part
- ☐ Users be trained in the proper use and care of the instruments
- ☐ Instruments be recalibrated after abuse, dropping, etc.
- ☐ Instruments be calibrated at specified intervals

Measurement Standards

To measure a process or product, standard measurement units are required to measure product characteristics and compare measurements against others. The United States uses English units while the rest of the world uses metric or SI (Système International d'Unités) units. SI units subsequently evolved into the metric system.

Measurement standards form a hierarchy. At the lowest level of the hierarchy are working standards. Measurement instruments used by production or quality inspectors are calibrated against these standards at regular intervals.

Working standards are then calibrated against a higher and more accurate standard, called a **reference standard.** A precision of 10 : 1 is usually required to transfer from one standard to another.

Reference standards are referred to the highest, most accurate set of standards, called primary reference standards. The National Institute of Standards and Technology (NIST), formerly National Bureau of Standards (NBS), maintains these standards. So when a measuring equipment is calibrated to a reference standard, it can be traced to an NIST standard.

Repeatability

Measuring instrument **repeatability** is as important as ensuring machine repeatability. Just as machines have to be monitored, measuring instruments are calibrated to control and assure measurement repeatability. The following three concepts are important to understanding measurement repeatability: resolution, accuracy, and precision.*

Resolution. Resolution is the sensitivity of a measuring instrument. Sensitivity is a function of the internal components and design of the instrument. For example, electronic instruments are more sensitive and accurate than mechanical instruments.

Accuracy. Accuracy is the extent to which the average of many measurements agrees with the standard being measured. The average of many measurements minus the standard value is the amount of error in the instrument. Poor accuracy usually results from a lack of calibration.

Precision. Precision is similar to repeatability. It is the ability to repeat a series of measurements and get the same value each time. Precision and accuracy are different concepts that are sometimes confused. For example, the accuracy of an electronic micrometer can be recalibrated to reduce its error. But recalibration usually cannot improve the micrometer's intrinsic precision (Figure 1.4).

---

* Datamyte, *DataMyte Handbook,* 1989, pp. 6.6–6.10.

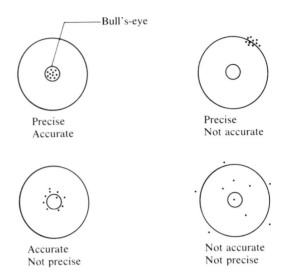

FIGURE 1.4
Precision and accuracy. From *Continuing Process Control and Process Capability Improvement,* © Ford Motor Company; used with permission.

Measurement Error

Measurement error is the difference between a measured value and the true value. Any difference is due to an error in accuracy or precision. Sources of error are environment, deformation, and human error.

Environment. Temperature, dust, and water are environmental factors that cause measurement error. For very precise measurements, environmental factors are controlled carefully. The measurement of critical dimensions to millionths of an inch is performed in a sealed area where the temperature, dust, and humidity are monitored and controlled. Products are tested in temperature-controlled environments in order to stabilize the temperature of the products to that of the area. Otherwise product dimensions will change in direct relationship to product temperature changes.

Deformation. **Deformation** is the second largest cause of error following temperature. Deformation is caused by compression of the measurement instrument against the measured part. If a gage's probe touches a part, it might bend or deform when pushed against the work piece.

Human Error. Human errors can be either intentional or unintentional. Examples of unintentional errors are loose clamps, poor connections, and improper handling or storage. This type of error is caused by poor instructions, lack of training, or inadequate procedures. Unintentional errors can be minimized through training personnel, understandable written procedures, and enlightened supervision.

Intentional errors usually result from poor attitudes rather than a lack of aptitude. Intentional errors might be eliminated through counseling and guidance or through a job transfer.

QUALITY PROBLEM SOLVING

The **systems approach** is a logical, systematic method for solving many types of quality problems. The value of this method lies in its ability to define a problem and to arrive at a solution through a logical process. The goal of the systems approach is not only to eliminate the **symptom,** put also to identify the **root cause** and to eliminate it as well. The systems approach follows the systematic methodology illustrated in Figure 1.5 and described in the following paragraphs.

Define Problem. The problem and its scope are first outlined. The problem is defined to ensure it is solvable. If it is not solvable with available organizational resources, including knowledge, personnel, financial and political, additional resources are requisitioned. Sometimes, even with additional resources a problem cannot be solved.

It is important that a problem be defined to determine whether it is solvable within the required time period and with available resources. A problem might become a major and costly project that taxes corporate resources and does not return the initial investment. In such a case, the entire CWQM program and any subsequent improvement projects might be jeopardized.

Modern industrial or service problems are complex. Usually one person does not have the skills or knowledge to solve them. So interdisciplinary teams are created to solve these problems. These teams consist of representatives from

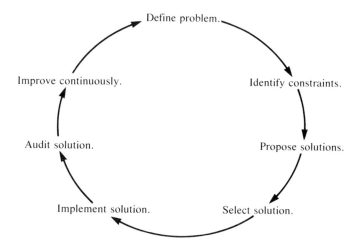

FIGURE 1.5
Systems approach.

the area where the problem is located as well as experts from peripheral areas who can supply information. For example, if a vendor is supplying a defective product, a team composed of personnel from purchasing, engineering, and manufacturing is formed to work with the vendor to improve product quality.

Identify Constraints. Constraints to solving the problem or implementing solutions are identified. Constraints are factors that might have to be anticipated and eliminated as problems arise. Common constraints are cost, time, size, regulations, environment, permits, and culture.

With a quality-related decision, cost is always a major constraint. Any product or service has to be cost-effective. However good a product or service, it will only be purchased if it is priced competitively.

Propose Solutions. The team brainstorms and proposes possible solutions. In the nonthreatening atmosphere of a brainstorming session, people are less constrained and are free to offer unconventional and novel ideas for problems that previously were thought to be unsolvable. Sometimes the most bizarre proposal turns out to be the most innovative solution.

Select Solution. Each proposed solution is evaluated in terms of its ability to overcome the constraints discussed and to achieve corporate objectives depending on selection criteria. Proposals are ranked and quantified based on corporate requirements. The solution that is most cost-effective, easiest to implement, generates the most revenue, or saves the most money is selected.

If the problem is a defective product or a deficient machine, the selected solution should attack the root cause of the problem. For example, if the identified cause of a failed pump is a defective seal, then replacing the seal would get the pump operating, but not prevent premature pump failure. The seal is not the root cause solution. A properly designed and implemented preventive maintenance program could have extended the life of the seal.

Implement Solution. Next, the solution is implemented so that problems do not recur. People in the area where the problem is located are responsible for its implementation. These people have a personal interest in its implementation and the elimination of the problem. These people also monitor implementation over time, so that the problem does not recur. Schedules are established, and project milestones are identified and tracked.

Audit Solution. The solution is finally monitored for effectiveness, cost, and reliability. If the solution does not solve the problem, further **corrective action** may be warranted. In corrective action, the team goes back to identifying the problem and repeating the whole cycle, always keeping in mind that the goal is continuous improvement. This forms a feedback loop, so that a solution is optimized and the problem does not recur.

Improve Continuously. Continuous improvement in a product or service has become a necessary element of an organization's survival in a global economy. In times past, if a firm had a quality or low-priced product, the firm probably dominated the market. There were several reasons for this. The number of firms producing the product was limited. There were also barriers for firms trying to enter the market, such as tariffs, cultural restrictions, cost, technical knowledge, or management skills.

Now there is no such thing as long-term ownership of a market. In a global economy, many of these constraints are gone, and if a market for a product or service evolves, an organization somewhere will emulate or improve on the existing product.

SPOTLIGHT

The American Society for Testing and Materials (ASTM), founded in 1898, is a scientific and technical organization that compiles technical specifications. Its mandate is "for the development of standards on characteristics and performance of materials, products, systems, and services; and the promotion of related knowledge."

ASTM develops standards through voluntary consensus of its technical committees. The committees are composed of technical and industry experts who represent the views of producers, users, and general-interest participants. ASTM has developed thousands of standards on many products, from steel piping, concrete, wood, paint, textiles, rubber, electronics, nuclear components, to testing.

ASTM published a standard for polyethylene (PE) plastic tubing (D2737-81). This tubing is used in plastic garden hose. The standard specifies criteria for classifying plastic tubing materials as well as requirements for material composition, workmanship standards, and physical dimensions. Also, the standard outlines specific requirements for sustained pressure, burst pressure, and environmental stress cracking tests.

A typical part of the standard specifies that the wall thickness and tolerance for 1.00-inch (PE 2305) plastic tubing shall be a minimum of 0.154 + 0.015 inch.

The workmanship element of the standard states: "The tubing shall be homogeneous throughout and free from visible cracks, holes, foreign inclusions, or other defects. The tubing shall be as uniform as commercially practicable in color, opacity, density, and other physical properties."*

* *1982 Annual Book of ASTM Standards,* Part 34, "Plastic Pipe and Building Products," pp. 372–378.

SPOTLIGHT

The U.S. government purchases many products, systems, supplies, and services. The government has developed a standard that is used by many large government agencies, such as the Department of Defense, to ensure that suppliers have a quality program.

The standard, commonly referred to as MIL-STD 9858A, requires that suppliers of critical or major products have an extensive and documented quality program, ensuring that procedures, processes, and products comply with specifications.

Military procurement practices have come under much scrutiny as a result of several major incidents involving flawed products, overpriced products, and catastrophic failures. Military hardware is more sophisticated and complex. As complexity rises, so does the potential of a flaw. Military products must be reliable and maintainable by soldiers with the equivalent of a high-school education. A major cause of the space shuttle Challenger disaster was the poor quality of supplied parts. A supplier was providing a flawed O-ring, and quality inspection at NASA facilities was inadequate.

The government is now mandating compliance with this quality standard up to third-tier suppliers of critical products. This means that the suppliers of the prime contractor's supplier must also comply with this document.

Military contract requirements for a missile system or a new jet fighter consist of many thousands of pages of drawings, documentation, and specifications. A jet fighter contract may be worth $20 billion over the life of the contract. Every element of a piece of military hardware is specified. The military wants to know that quality is understood, and specified by the contractor.

The government quality standard specifies that quality shall be pursued in design, development, fabrication, processing, assembly, inspection, test, maintenance, packaging, shipping, storage, and installation. In each of these areas, quality is monitored, controlled, measured, and documented.

The goal of the military quality standard is defect prevention. Prevention requires that all elements of the organization monitor and control their processes and products. All supplies and services within a supplier's plant and sources are controlled to assure conformance to contract requirements. If deviations arise, they are corrected immediately.

The standard states that effective management for quality is prescribed by the contractor. This means that quality personnel have sufficient, well-defined responsibility, authority, and organizational freedom to identify quality problems and initiate action that eliminates problems.

For example, the jet fighter will be modified many times through the life of the contract. Newer, lighter, corrosion-resistant material might be specified to make the fighter faster and more maneuverable. Each time there is a revision, changes are incorporated in the drawings, so that the drawings are always complete, current, and adequate. If drawings are not revised, a mistake will be replicated in all manufactured parts. It is very expensive to retrofit parts into an operational product.

The contractor is responsible for assuring that all procured supplies and services conform to contract requirements. This means suppliers must comply with the same requirements as the prime contractor.

The contractor must also assure that all machining, wiring, batching, shaping, and all basic production operations are accomplished under controlled conditions. Work instructions are documented as well as production equipment. Products may also require specialized testing or inspection. Inspection equipment is calibrated at prescribed intervals and is stored in secure and clean areas.

SUMMARY

- ☐ Quality is a dynamic concept that is continuously adapting to respond to changing customer requirements.
- ☐ Quality is defined in three ways: conformance to specifications, fitness for use, and satisfaction of customer expectations at a competitive price.
- ☐ Quality as a functional area has evolved through the following stages: inspection, quality control, quality assurance, and company-wide quality management. Company-wide quality management is also called total quality management.
- ☐ In company-wide quality management everyone in the organization is involved with the quality effort. Products and services from conception to delivery have a customer and quality focus.
- ☐ Quality loop identifies company-wide quality management activities from marketing and market research to disposal after use.
- ☐ Product brief distills and describes customer requirements to the organization. The brief includes performance characteristics, aesthetics, serviceability, packaging, price, and regulatory requirements.
- ☐ Engineering uses the product brief to translate vague customer requirements into a tangible product. Engineering considers customer requirements, cost, manufacturability, testability, and design quality when developing a product.
- ☐ Procurement is especially important in developing a quality product because 70% or more of the cost of a product involves outside suppliers of critical components.
- ☐ Production is responsible for creating reality out of engineering specifications. Production is a term that includes manufacturing, stores, assembly, fabrication, and inspection.
- ☐ A quality product is only as good as after-sales technical assistance and service. If a customer experiences poor after-sales service, the result could be loss of customer loyalty and subsequent sales.
- ☐ Prevention is preferred over inspection. Inspection results in loss of value because of scrap and rework. Prevention adds value to a product.
- ☐ Quality is only as good as the accuracy of the measuring instrument. If data are inaccurate, quality decisions will not be reliable.
- ☐ Quality problem solving should be a systematic process; otherwise problems will not be solved in an expeditious or cost-effective manner. The systems methodology consists of the following steps: define problem, identify constraints, propose solutions, select solution, implement solution, audit solution, and improve continuously.

KEY TERMS

accuracy   Extent to which the average of many measurements from a calibrated measuring instrument agrees with the standard being measured.

American National Standards Institute (ANSI)   American standards-making body.

American Society for Quality Control (ASQC)   Professional organization dealing with technical and managerial quality issues.

company-wide quality management (CWQM)   Program, attitudes, and culture that ensures product and service quality is pursued throughout the organization from conception to delivery.

corrective action   Action to eliminate symptomatic and root causes of quality-related problems.

deformation   Alteration of shape of measuring instrument against the measured part; major cause of measurement error.

inspection   Measuring of products against specifications.

International Standards Organization (ISO)   European standards-making body.

manufacturability   Ease of production, fabrication, or assembly, where no superfluous movement of machine, material, or people is involved.

precision   Ability to repeat a series of measurements and obtain the same value each time; similar to repeatability.

prevention Deterrence of defects, flaws, or non-conformances; opposite of inspection.

product brief Document communicating customer requirements to the organization, including performance characteristics, aesthetics, serviceability, packaging, price, and regulatory requirements.

product life cycle Sales pattern of a product, consisting of four stages: introduction, growth, maturity, and decline.

quality Depending on the perspective of the user: conformance to specifications; fitness for use; or satisfaction of customer wants, needs, and expectations at a competitive price.

quality assurance (QA) Process of ensuring that quality is pursued in organization.

quality control (QC) Control of quality through check, test, and verification of quality attributes in order to indicate conformance to customer requirements.

quality loop Combination of the following CWQM elements: marketing and market research; design/specification engineering and product design; procurement; process planning and development; production; inspection, testing, and examination; packing and storage; sales and distribution; installation and operation; technical assistance and maintenance; and disposal after use.

reference standard Primary standard that defines unit of measure.

reliability Long-term quality.

repeatability Ability to obtain the same value from a measuring instrument after repeated tests.

resolution Sensitivity of a measuring instrument.

root cause Prime cause of a problem such that, if removed, the problem will be eliminated.

specification Type of standard that precisely states a set of requirements to be satisfied by a material, product, system, or service; requirements may be numerical or visual; a test procedure shall be defined that determines whether the given requirements have been satisfied.

standard A specification, test method, definition, or practice that has been approved by a professional organization, industrial trade association, government body, or regulatory authority.

statistical process control Statistical method of prevention in which process variation is monitored and, if necessary, adjusted.

symptom Effect of some defect or flaw.

systems approach Systematic procedure for solving quality problems.

test Examination that determines the properties, composition, or performance of materials, products, systems, or services.

testability Ease of testing, measurement, and inspection.

QUESTIONS

1. What are important trends in quality?

2. What is quality? Discuss the significance of each definition.

3. Would you purchase a product regardless of price? Why?

4. What are the major elements of CWQM? Explain each.

5. What are possible conflicts of interest that can arise when quality reports to manufacturing?

6. What is a product brief?

7. How are customer expectations transformed into product requirements?

8. Discuss the problems of overly tight or loose specifications.

9. What requirements should be communicated to suppliers?

10. Discuss the importance of service quality compared to product quality.

11. What are the major elements of the problem-solving methodology discussed in this chapter?

12. What are differences between inspection and prevention?

2

Quality Statistics I

Statistics deals with the collection, analysis, presentation, and interpretation of quantitative data. The purpose of statistical analysis is to derive information from raw data in order to make intelligent decisions on quality.

Almost every element of an organization implementing a company-wide quality management (CWQM) program uses statistics to analyze problems, develop solutions, and improve processes. For example, marketing uses inferential statistics to determine the size and makeup of customer surveys. Engineering uses probabilistic methods to enhance the reliability of products. Purchasing uses statistical analysis to evaluate and monitor suppliers. Quality assurance uses statistics to determine the number of products that should be inspected. Manufacturing uses statistical process control to monitor production processes. Management uses graphic methods to display data.

STATISTICAL FUNDAMENTALS

Populations and Samples

Before we begin our study of statistics, we define several basic terms in order to develop the fundamentals of statistical quality analysis. The study of statistics starts with an understanding of populations and samples.

A **population** is a group of products, observations, individuals, or measurements. It can be a shipment, a lot, or a batch of products. It can be a data set of observations from an experiment. It can be workers in a machine shop, in a work area, or on a shift. Also, a population can consist of a group of measurements from a production process.

A **sample** is a small representative group of products or items that is selected from the population. A representative sample has the same relevant quality attributes and in the same proportion as the population. Several examples may help to illustrate these concepts.

An operator machines 150 parts to a specified dimension every hour. The machinist pulls five parts (sample) every hour from the 150 parts (population) and measures the dimension that was just machined. The operator then calculates the average value and the difference between the highest and lowest values of the five dimensions and plots these values on a statistical process control chart.

Quality control wants to evaluate the quality of an incoming shipment of products. A quality analyst does not check all products in the shipment (population), but pulls a small representative group (sample) from the shipment. If the shipment (population) is composed of 1% defective parts, the sample should hopefully also be 1% defective. The quality analyst infers information of the shipment based on the sample. Specifically, the analyst accepts or rejects the shipment based on the results of the inspection of the sample.

Selecting Samples. Why select a sample instead of studying the entire population of a product (Figure 2.1)? As a practical matter, checking 100% of a population is not effective; 100% inspection is costly because it requires additional personnel, time, equipment, and space. Studies of 100% inspection indicate that it is only 80% effective. Inspectors become hypnotized and tired of the monotonous inspection of similar products.

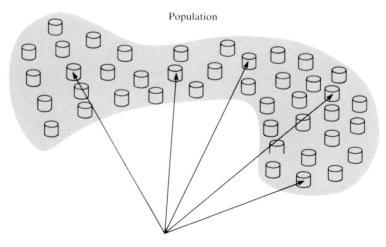

Population

FIGURE 2.1
Selecting a random, representative sample.

Sampling involves certain risks. Nonconforming products might be accepted or good products might be rejected; or products can be selected from a continuous process and might not represent the entire production of the product. Five products selected at 9:00 A.M. might not represent products being produced at 6:00 P.M. when an inexperienced operator is producing nonconforming parts or using nonconforming raw material.

Types of Statistics

In this chapter and the next we discuss two types of statistics: descriptive statistics and statistical inference. A third type of statistics, probability, is discussed in Chapter 5. **Descriptive statistics** is the study of organizing, summarizing, presenting, and displaying data. Descriptive quality data can be organized into meaningful patterns through arrays, frequency distributions, frequency polygons, Pareto diagrams, and cause and effect (C–E) diagrams.

Statistical inference involves deriving conclusions of a population based on characteristics of a sample. Statistical inference is also called inferential or inductive statistics. Marketing uses statistical inference techniques to determine the appropriate sample size of a market survey to evaluate a new product. If the sample group is too small, not representative of the population, or not selected randomly, the customer survey results will not provide accurate information about the population.

In some ways, **probability** is the reverse of statistical inference. Probability is the statistical study of a sample based on knowledge of the population. Specifically, a quality analyst uses probability to make statements about the likelihood of a sample having certain characteristics based on available information of the population. For example, a quality analyst can determine the probability of selecting one defective product out of a sample of four products, knowing the population is 99% conforming to specifications.

These three areas of statistics do no stand alone in the study of quality; they are integrated throughout this book. If you decide to continue your study of the technical aspects of quality, you will learn advanced statistics, including design of experiments, decision making, and multiple regression analysis.

RAW DATA

Quality Data

Statistical data obtained through a customer survey, product test, or inspection consist of raw, unorganized, and meaningless numbers. The data impart little information that can be used by management to make reliable decisions. Data must first be collected and organized into meaningful formats before they can be interpreted.

Raw data reflecting quality levels can be obtained by measuring product quality characteristics, counting defective products, measuring response times for delivering a service, and in numerous other ways. Data should serve as the basis for rational decisions and for subsequent management action. Action is doing something to eliminate the cause of a deviation to a specification or standard and hopefully eliminating its root cause. Action could be modifying a product to eliminate a flaw, improving a production line, training employees in serving the customer, or repackaging a product to enhance its marketability.

In this book we are concerned with data obtained through the measurement of a product quality attribute, a production process, or a service process. For example, measurement data are collected to understand dispersion in a product's dimension or performance. If the product does not meet specifications, measurement data can reveal the amount of deviation and the correction required to produce conforming products. Sometimes even adjusting a process cannot produce conforming parts. Adjusting a process might not eliminate the root cause of the defect, which can be a shipment of very hard raw material or chips in a machine's cutting tool. Or the tolerance spread of the specification might be too narrow for the process to make parts that satisfy the specification.

Important Questions

Raw data consist of numbers, which by themselves reveal little useful information. Raw data are the inputs to statistical processes, just as raw materials are the inputs to manufacturing processes. Raw materials are processed through many steps until a quality product is produced. Likewise, raw data are processed through many statistical steps until reliable information is produced.

Before raw data are analyzed through statistical techniques, they should be analyzed to ensure their usefulness and accuracy. If raw data are flawed, the output will also be flawed. The analogy of a computer is helpful. A computer is a machine that manipulates data. If the inputted data are suspect, the output will also be flawed—garbage in and garbage out.

Useful Data. Data are useful if they can be used to identify and eliminate the immediate symptom and root cause of a product defect or machine failure. On the other hand, data can accumulate and overwhelm production personnel. Some processes automatically inspect 100% of the parts coming down a manufacturing line, generating large amounts of information.

Whenever data are collected, asking basic questions helps to ensure that proper and accurate data are being analyzed. What is the scope of the quality study? What is the basic purpose of the study? Will acquired data answer the questions being asked? How will data be obtained? Who will perform the measurement and analysis?

Data Are Accurate and Precise. A frequent theme in this book is that quality can be determined, monitored, and controlled if input data are accurate and precise. Causes of inaccurate ,or imprecise data are measuring instruments, operators, methods, or raw material. A measurement instrument might not be calibrated. Different operators might have run a machine and commingled data. Methods of running the machine might vary from operator to operator. Raw material might show varying physical properties.

DESCRIPTIVE STATISTICS

Arrays

Once the quality of raw data has been ensured, data can be arranged into an array and then into a frequency distribution. An **array** arranges numbers in ascending or descending order. In the language of statistics, an array arranges observations in a data set in ascending or descending order. A **frequency distribution** arranges data into groupings that show the number of observations in different categories. An observation can be a measurement, individual, or item. We introduce an array and a frequency distribution through the following example.

A quality manager wants to develop a warranty for a 100-watt lightbulb. The quality manager selects a random, representative sample of 50 lightbulbs from a production lot and reliability tests them to failure. Reliabilities are measured in terms of hours to failure. The reliabilities of the 50 100-watt lightbulbs are as follows:

1983	2235	2414	2465	2510
2329	2414	2697	2567	2270
2321	2214	2130	2174	2353
2438	2356	2299	2238	2350
2450	2454	2452	2543	2544
2026	2237	2248	2643	2417
2326	2320	2293	2234	2343
2680	2565	2438	2564	2387
2027	2175	2346	2438	2652
2420	2355	2362	2146	2124

Constructing a Frequency Distribution

The process for constructing an array and a frequency distribution is fairly straightforward. A quality analyst must proceed as follows: collect the raw data, arrange the data into an array, group the data into classes, determine the class interval, construct the tally chart, and finally construct the frequency distribution.

Collect Raw Data. Raw data are collected through tests, measurements, or inspection. As we discussed, raw data must be useful and accurate, otherwise the reliability of the information can become questionable. Therefore, our quality manager was very careful in selecting lightbulbs that were representative of the production population. Also, reliability tests were stringently conducted under controlled conditions by trained personnel.

Arrange Data into Array. Data are next arranged or grouped into an array. An array creates order out of data; it can arrange numbers in descending or ascending order. An array displays the spread and the center of the numbers in the grouping. If an array has many numbers, a computer can sort these quickly.

The preceding raw data are arranged into an ascending array as follows:

1983	2235	2329	2414	2510
2026	2237	2343	2417	2544
2027	2238	2346	2420	2543
2124	2248	2350	2438	2564
2130	2270	2353	2438	2567
2146	2293	2355	2438	2565
2174	2299	2356	2450	2643
2175	2320	2362	2454	2652
2214	2321	2387	2452	2680
2234	2326	2414	2465	2697

Group into Classes. Data are then grouped into **classes,** also called **cells.** The number of classes is up to the analyst. A convenient method for determining the number of classes is to count the number of data points in the set and take the square root of this number. For example, if there are 100 numbers in the data set, the square root of 100 is 10, which is the number of classes. As a rule of thumb, if the square root is a decimal answer, the value is rounded off to the nearest whole number. If there are 200 numbers in the data set, there are 14 classes. It sometimes is necessary to add one or two more classes so that all the numbers in the data set can be incorporated into a class.

Determine Class Interval. Once the number of classes is determined, the class interval, sometimes called class width, is calculated by this formula:

$$\text{Class interval} = \frac{X_h - X_l}{\text{number of classes}}$$

where X_h = highest number in array
X_l = lowest number in array

There are 50 pieces of data in our example, and the square root of 50 is 7.07, or simply 7. Using the above formula, the class interval is calculated:

$$\text{Class interval} = \frac{2697 - 1983}{\text{number of classes}}$$

$$= \frac{2697 - 1983}{7} = 102$$

Since the spread of data is 714 hours, the quality manager believed that most of the data could be grouped into seven classes, each of which is 102 or, to make it simpler, 100 hours. For example, the first class interval contains all of the failure data from 1900 to 1999 hours inclusive. The next class is from 2000 to 2099 hours, and so on.

But with seven classes, four numbers would not be in any class interval. The four numbers are 2643, 2652, 2680, and 2697. Another class is therefore added from 2600 to 2699 hours to include these values.

Construct Tally Chart. The next step tallies, or adds, the numbers in each class interval. There is one tally in the 1900–1999-hour class interval, two tallies in the 2000–2099-hour class, and so on. A tally is sometimes called a class frequency.

An X is used to indicate an observation in a class interval. The tally can be arranged horizontally or vertically, depending on the analyst.

In this example, we tally data horizontally. The class intervals are listed on the left-hand side and the number of failures within each class interval is shown on the right. When completed, the tallies appear as follows:

Hours	Tally
1900–1999	X
2000–2099	XX
2100–2199	XXXXX
2200–2299	XXXXXXXXX
2300–2399	XXXXXXXXXXXX
2400–2499	XXXXXXXXXXX
2500–2599	XXXXXX
2600–2699	XXXX

Construct Frequency Distribution. The tallies can then be converted to a frequency distribution. The left side of the frequency distribution shows the class intervals and the right side the number of observations. The frequency distribution derived from the preceding tally sheet is as follows:

Hours	Number of Failures
1900–1999	1
2000–2099	2
2100–2199	5
2200–2299	9
2300–2399	12
2400–2499	11
2500–2599	6
2600–2699	4

What information can be obtained from this frequency distribution and tally sheet? The lowest failure rate of the 100-watt lightbulb is about 1900 hours and the highest is about 2690 hours. Most lightbulbs fail between 2200 and 2500 hours. The largest concentration of failures is between 2300 and 2399 hours.

In summary, an array and a frequency distribution allow the quality analyst to obtain information from raw data. The center of the distribution can be located as well as the spread of values around the central point.

A common problem is having a large data set and too many classes. If the number of classes is too large, recognizable patterns will not be detected. For example, an array of 1000 product dimensions would have 31 classes. This number can be lowered to compress the data into fewer classes.

Relative Frequency Distribution

Sometimes the relative frequency in each class is calculated instead of the actual number of observations. **Relative frequency distributions** are useful for forecasting the number of lightbulbs that would generally fail within the 2000–2099 class interval, assuming that the random sample represents the population of lightbulbs being produced.

To convert a frequency distribution to a relative frequency distribution, each class frequency is divided by the total number of frequencies. The total value of the relative frequencies is 1.0. Relative frequency is calculated by using the following formula:

$$\text{Relative frequency} = \frac{\text{number of observations in class interval}}{\text{total number of observations}}$$

Using the lightbulb example, the relative frequency of failures for the 2000–2099 class interval is

$$\text{Relative frequency} = \frac{2}{50} = 0.04$$
$$= 4\%$$

The relative frequency in each class interval of the lightbulb reliabilities is as follows:

Hours	Number of Failures	Relative Frequency
1900–1999	1	1/50 = 0.02
2000–2099	2	2/50 = 0.04
2100–2199	5	5/50 = 0.10
2200–2299	9	9/50 = 0.18
2300–2399	12	12/50 = 0.24
2400–2499	11	11/50 = 0.22
2500–2599	6	6/50 = 0.12
2600–2699	4	4/50 = 0.08
		1.00

In the next section we show how frequency distributions can be graphed. We explain two types of **bar charts** that are derived from frequency distributions, histograms and frequency polygons.

FREQUENCY DISTRIBUTION GRAPHS

Histograms

A frequency distribution graph, such as a frequency **histogram,** illustrates the results obtained from an array or tally sheet. A graph communicates information more easily than do numerical analyses. By interpreting a graph, a quality analyst can count the data in each class interval and determine the center and the spread of the distribution data.

A frequency histogram is a bar graph that shows the results of the frequency distribution analysis. A relative frequency histogram shows the results of the relative frequency distribution analysis. Either type of histogram compresses raw data into logical groupings or categories (see Figures 2.2 and 2.3).

To explain the charts, we again use the lightbulb example. In a typical histogram, the quality characteristic is listed on the horizontal axis. For example, the class interval 2300–2399 is scaled on the horizontal axis. Class frequencies, the frequency of occurrence, are scaled on the vertical axis. The height of each bar corresponds to the number of measurements in that class interval.

The classes in the frequency distribution of lightbulb failures start at 1900–1999 hours and end at 2600–2699 hours. Each class interval has a lower class limit and an upper class limit. Usually the lower limit of the first class is below the first number in the array and the upper limit of the final class is above the last number in the array (see Figure 2.4).

The midpoint of each class interval becomes the center of each bar in the histogram; it is halfway between the class limits. The midpoint is calculated by adding the upper and lower class limits and dividing by 2. For example, the midpoint of the 2000–2099 class interval is

$$\text{Midpoint} = \frac{2000 + 2099}{2} = 2049.5$$
$$\sim 2050$$

Reliabilities in our example are rounded to the nearest whole number. So if a reliability is 2099.6 hours, it is rounded to 2100 hours, which would put it into the

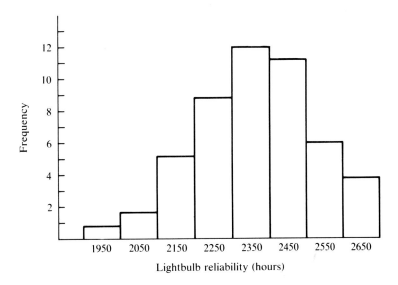

FIGURE 2.2
Frequency histogram.

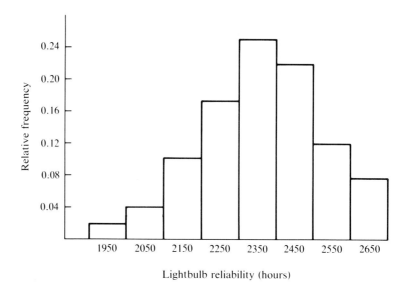

FIGURE 2.3
Relative frequency histogram.

next higher class interval. For simplicity, the midpoint of the interval is also rounded to the next highest whole number, which is 2050.

The horizontal axis is divided into the number of classes determined in the previous step. Each class interval has its midpoint value written in the middle of the class interval. The number of times that a measurement occurs is entered on

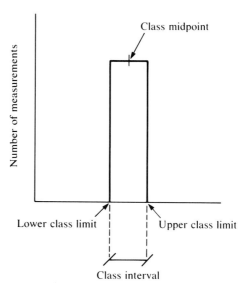

FIGURE 2.4
Structure of frequency distribution class.

the vertical axis. Once the number of measurements has been determined, bars are drawn to that height and as wide as the interval from the lower class limit to the upper class limit. The bar height equals the number of measurements in the class interval.

Frequency Polygons

Frequency polygons are another way of displaying frequency distribution data. A frequency polygon is constructed like a histogram. It is a line graph that connects the midpoints of the class intervals of all the bars in the histogram.

Frequency is shown on the vertical axis and classes are on the horizontal axis. However, instead of drawing a set of bar graphs, midpoints of each class interval are connected by straight lines. Again, the class midpoint is midway between the class limits. The result is a graph that is shaped like a polygon (see Figure 2.5).

A relative frequency polygon graphs the relative frequencies of each class interval instead of the actual number of observations. The relative frequency polygon of the same data has the same shape as the frequency polygon, except that the vertical scale corresponds to the relative frequency instead of the actual number of observations.

Histogram versus Polygon

Histograms and polygons both display frequency distribution data. Each has its distinct advantages. A histogram shows each separate class interval of the distri-

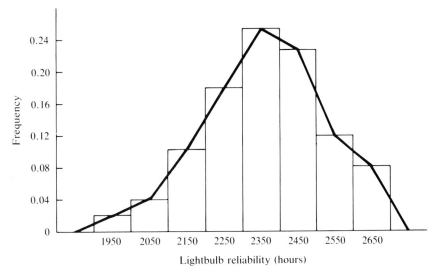

FIGURE 2.5
Frequency polygon.

bution, and the area of each bar corresponds to the total number of observations in the class interval. A frequency polygon is a line graph that is easily understood. As the number of classes and the number of observations increase, a frequency polygon will approximate the shape of a smooth curve. The next section describes four characteristics used to visually decipher smooth frequency distributions.

INTERPRETATION OF GRAPHS

Frequency Distribution Parameters

Frequency distributions display data in common forms and shapes. Numbers have a tendency to cluster together and show similar patterns. These patterns can be identified, measured, and analyzed. In this section we identify common distribution patterns, and in the next chapter we show how they are measured. The four major characteristics of a frequency distribution are central tendency, dispersion, skewness, and kurtosis.

Central Tendency. **Central tendency** is the characteristic that locates the middle of a distribution. Measures of central tendency are sometimes called measures of location. Figure 2.6 shows symmetrical curves with different central tendencies. Curves *A* and *B* are different symmetrical curves, but have the same central tendency. Curves *B* and *C* are similar symmetrical curves, but have different central tendencies.

Dispersion. **Dispersion** is the characteristic that indicates the amount of spread in the data. Dispersion is also called variation. Figure 2.7 shows two curves with different spreads. As can be seen, curves *A* and *B* have the same central tendency, but curve *A* has a wider dispersion than curve. *B*.

Skewness. **Skewness** is the characteristic that indicates the amount of distortion in a symmetrical curve. A symmetrical curve has the same shape to the left and

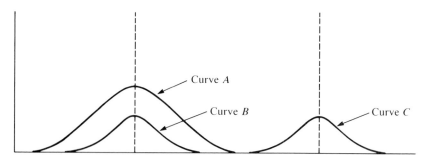

FIGURE 2.6
Central tendency characteristic.

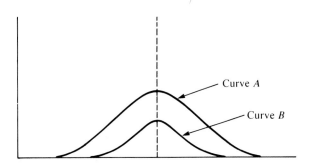

FIGURE 2.7
Dispersion characteristic.

right of the centerline. The two sides of a symmetrical curve are mirror images of each other [See Figure 2.8(a)].

In Figure 2.8(b) curves *A* and *B* are skewed right and left, respectively. These curves are skewed because data are clustered right and left of the distribution's centerline. Curve *A* is skewed to the right because most values are clustered on the right side of the distribution. Curve *B* is skewed to the left since most values are clustered on the left side.

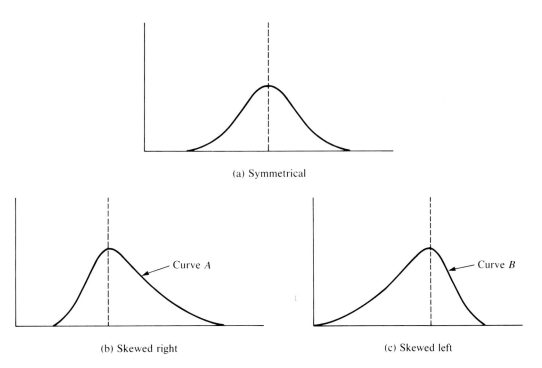

(a) Symmetrical

(b) Skewed right   (c) Skewed left

FIGURE 2.8
Skewness characteristic.

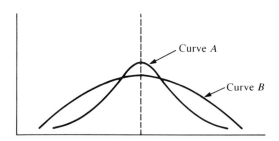

FIGURE 2.9
Kurtosis characteristic.

Kurtosis. Kurtosis is the characteristic that describes the peak in a distribution. It is a relative measure used to compare the peak of one distribution against another. A higher kurtosis value means a higher peak of reactive frequency, not just more data. In Figure 2.9 curve *A* has a higher kurtosis than curve *B* because its peak is higher.

There are three main classifications of kurtosis: mesokurtic, leptokurtic, and platykurtic (Figure 2.10). **Platykurtic** means that the distribution curve is broad and flat compared to the **leptokurtic** curve, which has a slender and high peak. In between, the **mesokurtic** curve has a peak less than the leptokurtic but higher than the platykurtic curve.

In the next two sections we discuss two special types of curves used in quality analysis, the Pareto chart and the cause-and-effect diagram. They are important graphs that we refer to throughout this book.

Pareto Diagrams

A **Pareto diagram** is a special form of histogram. It is a combination line and bar graph. A Pareto diagram, named after the Italian economist Vilfredo Pareto, is a

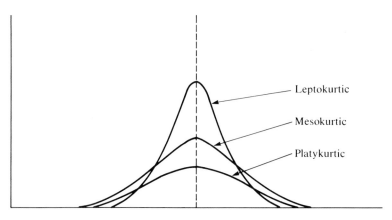

FIGURE 2.10
Curves showing main classifications of kurtosis.

visual method for identifying significant problems. These are arranged in a bar graph by their relative importance (Figure 2.11).

In the 1940s, Dr. Joseph Juran, an eminent quality expert, developed the Pareto Principle in quality. Dr. Juran said that the "vital few" problems are more important to solve than the "trivial many." This tendency, sometimes called the 80–20 rule, states that 20% of the problems (vital few) result in 80% of the nonconformances.

A Pareto diagram is not a problem-solving tool, but a tool for analysis. It is used to determine which problems to solve and in what order. It is wiser to identify and eliminate major problems, which leads to quicker cost savings. For example, by using the chart, it is easy to see which suppliers provide the lowest quality or which machines generate the highest level of scrap.

Used with a Cause-and-Effect Diagram. Figure 2.12 illustrates how a Pareto diagram is used with a **cause-and-effect (C–E) diagram.** The most significant problem identified in the Pareto diagram [Figure 2.12(a)] becomes the effect in a C–E diagram [Figure 2.12(b)]. This way scarce corporate resources are directed at activities that return the most benefits. As causes are identified, they are prioritized in another Pareto chart [Figure 2.12(c)].

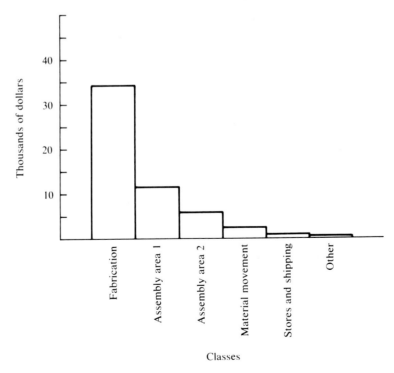

FIGURE 2.11
Sample Pareto diagram.

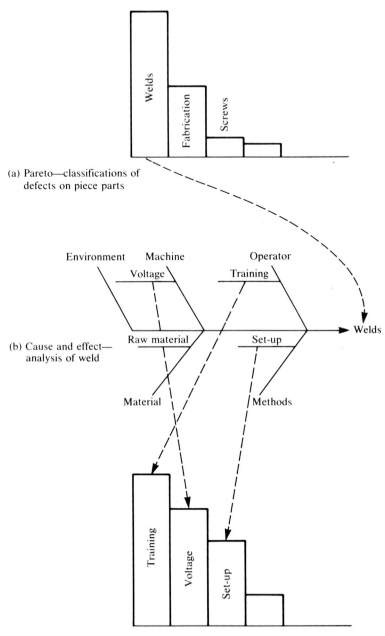

(a) Pareto—classifications of
    defects on piece parts

(b) Cause and effect—
    analysis of weld

(c) Pareto—relative
    importance of causes

FIGURE 2.12
Using Pareto and cause-and-effect diagrams. (a) Pareto: classifying defects on part. (b)
C–E: analyzing weld. (c) Pareto: prioritizing causes.

Once a problem has been eliminated, the next significant problem on the Pareto diagram becomes the effect on another C–E chart. The process continues until the vital few problems have been eliminated.

Pareto Chart Construction. The mechanics of constructing a Pareto chart are similar to that of constructing a histogram. First, categories are defined. They are derived from analyzing cost, defects, or defective data. These categories go on the horizontal axis. The vertical axis is scaled in proportion to the size, cost, or number of observations associated with each category. The category with the highest value or frequency goes on the left-hand side, next to the vertical axis. Then the categories are arranged in descending order, from left to right. The vertical and horizontal axes are divided into logical increments based on the frequency of occurrence and the number of categories. Finally, vertical and horizontal bars are drawn and the categories are listed below the horizontal axis.

Cause-and-Effect Diagrams

How are causes for nonconformances and defective products identified? The C–E diagram is a simple visual method for understanding cause-and-effect relationships. This diagram is also called the **Ishikawa diagram** after its inventor, Kaoru Ishikawa.

It has been said that the study of quality is the study of the dispersion, or variation, of data. The C–E diagram identifies the effect and lists the major causes of dispersion. If the cause of dispersion can be identified, the effect, which way be a quality defect, can be eliminated.

The C–E diagram is sometimes called a **fishbone diagram** because of its shape. On the right side of the drawing the effect, or problem, is written. Then a long horizontal arrow is drawn pointing to the effect. Coming off this line, or backbone, is the skeleton, which identifies the categories of causes. Smaller bones are added for each cause. When the diagram is completed, the causes are connected by arrows to show their relationships to each other and to the effect.

Categories depend on the nature of the problem. The major categories of a C–E diagram used to analyze the dispersion in manufactured products are material, operator, machine, method, and environment.

Material. This category includes raw or partially finished material used to make a product. Material may not be uniform because of different chemical, thermal, or physical properties.

Material uniformity is maintained by evaluating and selecting suppliers carefully. If the quality of the supplied material starts to vary, the vendor is notified of the problem and is responsible for correcting it.

Operator. This category includes men and women who cause dispersion using different methods or having different abilities. This type of dispersion can be minimized through training and proper procedures.

Machine. This broad category includes fabrication, assembly, test, measuring, and robotic or handling equipment. Dispersion is caused by machine wear, set-up, vibration, misuse, or lack of machine capability.

Machine dispersion of dimensional data can be minimized through preventive maintenance and proper use. Measurement instrument dispersion can be controlled through calibration, handling, and storage.

Method. Methods are instructions, procedures, or software used to perform some task. Methods to manufacture the same part can differ between shifts, plants, or processes.

Uniformity can be controlled by ensuring that all operators are trained extensively in machine use, material handling, and measurement. Also, procedures and work instructions should be uniform between shifts, similar work centers, and plants.

Environment. The first four factors are the primary categories found on C–E diagrams used to evaluate manufacturing problems. Environment is sometimes added. It includes factors such as internal politics, federal regulations, state regulations, temperature, humidity, and dust.

These factors can be controlled through compliance with appropriate state and federal regulations. Plant factors are controlled through careful monitoring and control of the plant's environmental conditions. For example, heating, ventilating, and air-conditioning systems can be set to adjust automatically for environmental swings.

Procedure for Drawing C–E Diagram. The procedure for drawing a C–E diagram, illustrated in Figure 2.13, is simple and straightforward. Its simplicity hides its ability to decipher complex situations. Drawing a C–E diagram includes the following steps: identify effect, identify causes, construct diagram, and analyze diagram.

Identify Effect. First the effect is identified and written on the right side of the C–E diagram. The effect might be a problem listed on a Pareto diagram. The effect should not be confused with a symptom. To avoid this confusion, operations specialists must analyze operations to determine the effect.

Identify Causes. Next, major categories of causes are identified. These usually consist of material, operator, machinery, method, and environment. Other categories may be used, depending on the effect being investigated. A group examining a service problem may divide the diagram into operator, method, software, computer, and environment.

Under each major category, minor causes are identified through brainstorming or similar techniques. Brainstorming is a method for generating a large number of ideas. Ideas are proposed by members of the group and written on a flip chart. Each member has an equal opportunity to express ideas. Open and candid discussion is encouraged.

Brainstorming is a proven and effective method for generating ideas in a nonthreatening atmosphere. Most employees are aware of problems that influence the quality of their work and will contribute to enhance the quality of their worklife.

Construct Diagram. Once ideas have been solicited, the causes are separated and listed under the main categories of the C–E diagram. A facilitator helps

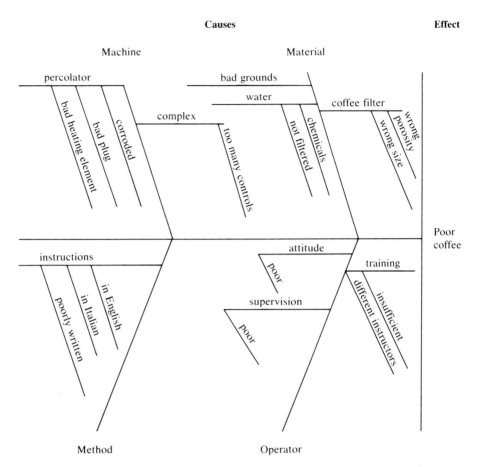

FIGURE 2.13
Sample cause-and-effect diagram.

generate ideas, organizes them, and lists them on a blackboard or flip chart. Causes are entered on the chart as bones on the skeleton of a fish. Causes can be identified that are subcauses of other causes. These are drawn as smaller bones or branches. The process continues until all of the causes are written down.

Analyze Diagram. Once a C–E diagram has been drawn, the work group votes on the causes it thinks are most important. Another Pareto chart can be constructed that prioritizes these causes. For each cause, the following questions should be asked:

☐ Is the cause a root or a secondary cause?
☐ Can solutions be identified to eliminate the root cause?
☐ Can solutions be implemented to solve the problem?
☐ Is the solution cost-effective?

SPOTLIGHT

An inspector is inspecting the surface finish of a product. Each day 500 products are produced and 100% inspected. The inspector makes an accept/reject decision based on a visual examination comparing the product's surface finish against an acceptable sample. The surface finish is a cosmetic, decorative quality characteristic of the product that if blemished affects the product's marketability. At the end of the day, the inspector counts the number of defective products. The following are results of 30 days of inspection:

0	1	1	2	2	3
0	2	4	2	3	3
1	3	3	2	3	4
0	1	2	3	4	3
4	4	5	6	5	6

Presented in this form, the data are not understandable and not useful. Management does not know how many defects are normal to the operation or when red flags should be raised to warn of an abnormal situation.

Tally Sheet

If these data are tabulated, processed, and grouped into an understandable form, then more information is revealed. Figure 2.14 shows a tally sheet containing the inspection results.

The frequency distribution is illustrated in Figure 2.15. It shows the dispersion in defective products. The distribution is centered at 3 defective parts and is shaped like a bell. The distribution spread is 6 defectives per day, and it looks as if the average number of defectives per day is a little less than 3. Thus the organization has a quality benchmark upon which it can improve.

Additional Information

What happens if the inspector adds more information to the diagram? By tracking the daily number of defectives for six weeks, it is seen that the quality is deteriorating over time and that the greatest number of defects occur on Fridays. More information is now available to identify and eliminate the cause of the problem so that defects do not recur.

Number defective	Tabulation	Frequency
0	///	3
1	////	4
2	⊔⊦⊦ /	6
3	⊔⊦⊦ ///	8
4	⊔⊦⊦	5
5	//	2
6	//	2

FIGURE 2.14
Tally sheet for rejected products over 30-day period.

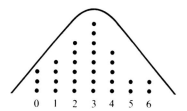

FIGURE 2.15
Frequency distribution for rejected products over 30-day period.

	Wk1	Wk2	Wk3	Wk4	Wk5	Wk6
Mon	0	1	1	2	2	3
Tue	0	2	4	2	3	3
Wed	1	3	3	2	3	4
Thu	0	1	2	3	4	3
Fri	4	4	5	6	5	6
Total	5	11	15	15	17	19
Cumulative	5	16	31	46	63	82

Trend Chart

Further information can be gleaned by constructing a trend chart. A **trend chart** displays information over time. The chart in Figure 2.16 shows that the surface finish is deteriorating over time, that the trend is accelerating, and that the total number of defectives each week is increasing, as indicated by the trend line becoming increasingly steeper.

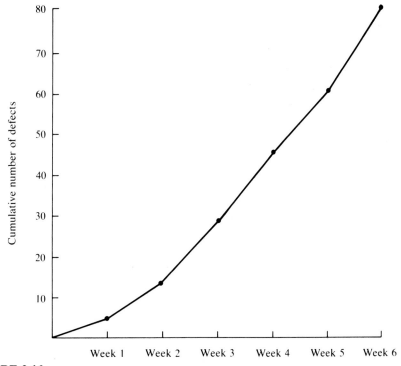

FIGURE 2.16
Trend chart showing product deterioration.

SUMMARY

☐ Statistics deals with collecting, analyzing, interpreting, and acting upon quantitative data.

☐ Almost every element of an organization uses statistics in quality analysis.

☐ A population is a group of items. A sample is a subset of the population.

☐ Care must be used in pulling a sample. A population should be homogeneous and the sample should be representative of the population.

☐ There are three types of statistics: descriptive, inferential, and probability. Descriptive statistics is the study of organizing, summarizing, and displaying data. Statistical inference derives information of a population based on the characteristics of a sample. Probability is reaching conclusions about a sample by studying the population.

☐ Collected data should be useful and accurate. Statistics processes data so that reliable information can be generated and action initiated. The action can be a management decision to dedicate funds for an improvement project or to initiate corrective action.

☐ When data are arranged in ascending or descending order, an array is generated. An array might indicate information through the manner in which it is presented. Patterns might develop. The center and the spread of the data become known.

☐ If data are divided into classes and graphed, additional information might be generated. A histogram is a simple bar graph where each bar is the class interval and the height is equal to the number of items falling within each class width.

☐ A relative frequency histogram and a frequency polygon can be generated from the histogram. A relative frequency histogram shows the relative frequency in each class interval instead of the actual number of observations. A frequency polygon is generated when the center points of the bars in a histogram are connected with a straight line.

☐ Data can also be organized into a frequency distribution.

☐ Frequency distributions can be characterized by four parameters: location, shape, dispersion, and kurtosis.

☐ A Pareto diagram, also called an 80–20 diagram, is used to isolate the major causes of a quality problem. The diagram, named after its inventor, suggests that 20% of the causes result in 80% of the losses.

☐ Two important specialized graphs used in quality are the Ishikawa diagram and the Pareto diagram. Ishikawa, also called a cause-and-effect diagram, shows the effect of a quality problem and its probable causes. The cause-and-effect diagram is divided into five major causes: material, operator, machine, method, and environment.

KEY TERMS

array   Arrangement of raw data in ascending or descending order.

bar charts   Graphic representation of the magnitude of data for different categories.

cause-and-effect (C–E) diagram   Diagram showing possible causes and effects of a quality problem; it is divided into five main categories: material, operator, machine, method, and environment.

cell   Same as class.

central tendency   Measure of location on distribution graph.

class   Range of values in frequency distribution.

descriptive statistics   Methods of organizing, summarizing, presenting, and displaying numerical data.

dispersion   Measure of spread on distribution graph; same as variation.

fishbone diagram   Diagram so named because of its resemblance to a fish skeleton; same as cause-and-effect diagram.

frequency distribution   Summary of a set of data in tabular form that shows the number of data points in each class.

frequency polygram   Line graph that connects

midpoints of class intervals of all bars in a histogram.

histogram   Bar graph of a set of data; the height of the bars is proportional to the number of items falling into each class.

Ishikawa diagram   Diagram developed by Kaoru Ishikawa; also called cause-and-effect diagram.

kurtosis   Peak level in a distribution.

leptokurtic   High-peaked distribution.

mesokurtic   Medium-peaked distribution.

Pareto diagram   Combination line and bar graph.

platykurtic   Flat-peaked distribution.

population   Group of items about which the quality analyst wants to draw conclusions.

probability   Area of statistics where a quality analyst selects a sample from a population and reaches conclusions about the sample, knowing characteristics about the population.

relative frequency distribution   Distribution that shows for each class category relative frequencies instead of number of observations.

sample   Group of items or products that are a subset of a population.

skewness   Measure of the amount of distortion from a symmetrical distribution curve.

statistical inference   Area of statistics where a quality analyst selects a sample from a population and reaches conclusions about the population knowing characteristics of the sample.

statistics   Techniques for compiling, analyzing, presenting, and interpreting data.

trend chart   Display of data over time.

QUESTIONS

1. What are the advantages of a frequency distribution compared to an array?

2. Explain how an array is constructed.

3. What are a class and a class interval?

4. Raw data are like raw material. Explain this analogy.

5. What are the steps in constructing a histogram?

6. As the numbers of observations and classes increase, what happens to the shape of the distribution? Explain.

7. When constructing a frequency distribution, the number of classes depends on what criteria?

8. What are the advantages of a cumulative frequency distribution?

9. Is a Pareto chart a problem-solving tool? Discuss.

10. What is the significance of the Pareto principle as it applies to quality?

11. Explain how to construct a cause-and-effect diagram.

12. What are five major causes of variation in a manufacturing process? Explain each.

13. How is a cause-and-effect diagram used with a Pareto chart?

PROBLEMS

1. Raw data from tensile tests of steel specimens are shown below. In this test, a piece of steel is placed between the jaws of a machine and pulled like taffy until the specimen fails. A sample is tested from 30 batches. Data are in units of thousands of pounds per square inch. Construct a histogram with a class interval of 2000 pounds per square inch.

60.4	57.8	61.2	60.4	58.2	59.1
59.3	56.3	59.7	62.1	65.5	55.1
60.3	59.2	56.9	58.3	62.4	62.6
60.4	61.2	59.2	57.2	64.5	65.8
56.9	58.8	62.5	64.1	61.5	63.2

2. Construct the following with intervals of one unit:
   a. Array
   b. Histogram
   c. Relative frequency histogram
   d. Frequency polygon

8	10	12	8	13	9	7	8	8	9
8	10	10	9	10	7	12	8	9	8
11	9	13	8	8	8	9	7	9	9
10	8	10	8	7	8	10	10	9	11
9	11	8	7	6	9	11	8	11	9

3. Using the data of Problem 2, construct the following with intervals of two units:
   a. Array
   b. Histogram
   c. Relative frequency histogram
   d. Frequency polygon
   What can you conclude if the intervals are too large?

4. The data are service delivery times in hours of an industrial delivery service. Construct the following using a class interval of 6 hours:
   a. Array
   b. Histogram
   c. Relative frequency histogram
   d. Frequency polygon

24	42	12	42	18	28	44	18	50	12
6	22	14	20	10	12	24	28	20	26

5. The following are delivery times in minutes for a pizza delivery service. Construct a frequency polygon.

56	4	16	34	48	14	40	52	26	18
30	26	38	46	44	10	30	24	56	40

6. A screw machine does not function. If you do not know the function of this machine, do your best to construct a cause-and-effect diagram for this problem.

7. A pencil sharpener does not work. Construct a cause-and-effect diagram for this problem.

3

Quality Statistics II

This chapter discusses measures of central tendency and dispersion, which are useful for describing frequency distributions. The four measures of central tendency are mean, weighted mean, median, and mode. The three measures of dispersion are range, variance, and standard deviation.

Using these concepts, the "normal" frequency distribution can be introduced. This distribution is often found in business and is characterized by a bell shape. If specification limits are displayed with a distribution, it is possible to calculate the amount of products that are outside the specification limits.

MEASURES OF CENTRAL TENDENCY

A **measure of central tendency** indicates the central location of a frequency distribution. We examine the following four measures of central tendency: mean, weighted mean, median, and mode.

Mean. The **mean** is the arithmetic average of a group of numbers. It is the most common indicator of central tendency. The arithmetic mean is calculated by adding all the values in a group and dividing by their number. In statistics, the values are called observations or elements, and the group is called a data set.

In order to obtain equations for the mean, we need to develop simple mathematical expressions. The equations for a sample mean and a population mean express the same concept of average but use different notations. The mean of a sample consisting of n elements is denoted by $\overline{X}$. The mean of the population consisting of N elements is denoted by μ (Greek letter mu).

The equations also use a new symbol, Σ, called the summation sign. Σ means that all the elements in a group are summed. The equations for the means of a population and a sample are therefore

$$\text{Population mean} = \mu = \frac{\Sigma X}{N}$$

$$\text{Sample mean} = \overline{X} = \frac{\Sigma X}{n}$$

where μ = population mean
$\overline{X}$ = sample mean
N = number of elements in population
n = number of elements in sample
Σ = summation

The calculation of population and sample means is straightforward. For example, a machine operator cuts 150 pieces of bar stock (round pieces of raw material) to a 3.000-inch specified dimension each hour. The operator pulls a sample of five pieces each hour and measures each length. The 150 pieces can be considered the population and the five pieces the sample. The sample mean is calculated as follows:

$$\text{Sample mean} = \frac{3.003 + 3.002 + 2.998 + 2.999 + 3.003}{5}$$

$$= 3.001 \text{ inches}$$

Intuitively, this sample mean seems correct because 3.001 is located in the middle of the data set spread.

To calculate the population mean, all of the measurements would be totaled and the sum divided by 150, the number of observations in the population.

The arithmetic mean is useful for expressing the central tendency of a frequency distribution. It uses all of the observations or measurements in a data set. Also, most people are accustomed to hearing about and using averages. On Sunday afternoons, football announcers talk about the average passing or running yardage of a football player. Economists discuss the state of the economy in terms of average income or average economic growth. The mean can be calculated easily and many calculators have a special function that simplifies the calculation.

However, the mean has a disadvantage as a statistical measure. If one number is especially large in a data set, it can distort the mean value. In the preceding example, if the last value in the data set is changed from 3.003 to 3.048, the sample mean changes also:

$$\text{Sample mean} = \frac{3.003 + 3.002 + 2.998 + 2.999 + 3.048}{5}$$

$$= 3.010$$

This average is distorted by one value, 3.048. Intuitively, this value does not seem to be located in the middle of the data set. Most of the numbers are clustered on one side, and there is another number far to the right on the number line (Figure 3.1). This distortion can be used to the advantage of the operator, who recognizes this as an abnormal reading. The operator finds the cause of the high reading and tries to eliminate its cause through corrective action.

Weighted Mean. The **weighted mean** is another measure of central tendency. The weighted mean concept is similar to the mean concept, except that it considers the importance of each value in relation to the whole data set.

This can be illustrated by an example. A company produces a product in three grades, "good," "better," and "best." The difference in the three grades is the type of raw material, fit of mating parts, and appearance. The best product requires more attention to detail and lasts longer than the good product. The three grades are priced differently to reflect additional costs in producing the products. The company wants to know the mean cost of material.

Type of Product	Cost of Material	Number Sold
Good	$10.00	1200
Better	$12.00	600
Best	$20.00	200

If we wanted to know the mean cost of material, we would use the mean formula:

$$\overline{X} = \frac{\Sigma X}{n}$$

$$= \frac{\$10 + \$12 + \$20}{3} = \$14.00$$

The problem with this calculation is that it does not consider the number of products sold. The weighted mean method considers the number of products sold by weighing the cost of each grade of product by the number of products sold.

For example, 2000 products were sold. Of the total number, 1200/2000, or 3/5, were of the good grade; 600/2000, or 3/10, were of the better grade, and 200/

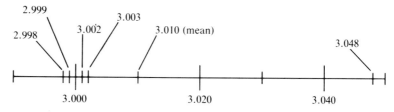

FIGURE 3.1
Number line.

2000, or 1/10, were of the best grade. Using these fractions as our weighted values for each of the grades of material, the weighted mean cost can be calculated:

$$\text{Weighted mean} = \left(\frac{3}{5} \times \$10.00\right) + \left(\frac{3}{10} \times \$12.00\right) + \left(\frac{1}{10} \times \$20.00\right)$$

$$= \$11.60$$

The weighted mean gives a value that is smaller than the simple mean because it considers the large volume of the cheaper, good products sold. The weighted mean equation is expressed as follows:

$$\overline{X} = \frac{w_1 X_1 + w_2 X_2 + \cdots + w_k X_k}{w_1 + w_2 + \cdots + w_k}$$

where $\overline{X}$ = weighted mean
w_1 = first weighing factor
w_2 = second weighing factor
w_k = kth weighing factor
X_1 = first number
X_2 = second number
X_k = kth number

This equation provides the same answer as the one we derived logically:

$$\overline{X} = \frac{\left(\frac{3}{5} \times 10\right) + \left(\frac{3}{10} \times 12\right) + \left(\frac{1}{10} \times 20\right)}{\frac{3}{5} + \frac{3}{10} + \frac{1}{10}}$$

$$= \$11.60$$

Median. The **median** is another measure of central tendency. It is the value that is the central number in an array of numbers. The median number separates the array so that half of the numbers are above and half are below the median number.

If the number of elements in an array is an odd number, the median is exactly in the middle. For example, if there are seven numbers in the array, the median is the fourth number because it is exactly in the middle of the array. If there is an even number of elements in the array, the median is the average of the two middle numbers. For example, if there are 10 numbers, the median is the average of the fifth and sixth numbers in the string.

Item in Array	1	2	3	4	5	6	7	8
	5	6	7	8	10	17	19	
				(median = 8)				
	2.3	4.8	6.9	9.4	9.8	12.3	15.1	19
				(median = 9.6)				

If there is an odd number of data in the array, the median is not a value that is calculated, the median is located. If there is an even number of data in the array, the median is calculated because it is the average of the two middle numbers.

The median can be located by using the following formula:

$$\text{Median} = \frac{n + 1}{2}$$

where n is the number of elements in the array.

This formula can be used if there is an odd or an even number of elements in the array. If there are seven numbers in the array, this formula locates the median as being the fourth item. If there are eight numbers in the array, the formula locates the median as being 4.5, or halfway between the two middle numbers.

The median offers several advantages over the average. Most importantly, the median is not influenced by values that are too large or too small. It can be visualized easily because it is the middle number in an array.

Mode. **Mode** is another measure of central tendency. The mode is the number that occurs most often in a group of numbers. The mode is not calculated. It is a number that can be visualized easily through examples.

In the following examples, the mode is readily identified.

Item in Array	1	2	3	4	5	6	7	8
	2	4	5	6	8	9	9 (mode)	
	1.2 (mode)	1.2	3	4	8	12	14	13.4

EXAMPLE The array shown lists the lightbulb reliabilities from Chapter 2:

1983	2235	2329	2414	2510
2026	2237	2343	2417	2544
2027	2238	2346	2420	2543
2124	2248	2350	2438	2564
2130	2270	2353	2438	2567
2146	2293	2355	2438	2565
2174	2299	2356	2450	2643
2175	2320	2362	2454	2652
2214	2321	2387	2452	2680
2234	2326	2414	2465	2697

The mean value is calculated by adding all of the reliabilities and dividing by 50. The mode is located. The median is first located in the center of the array; since there is an even number of data, the two middle numbers are averaged.

The mean, median, and mode are listed below:

Mean = 2359.3

Mode = 2438

Median = 2354

Locating the Mean, Median, and Mode

The mode sometimes can be understood by comparing it to the median and the mean in illustrations of smooth curve distributions. Figure 3.2 shows three distributions: symmetrical, skewed right, and skewed left. In each distribution the mean, the median, and the mode are indicated.

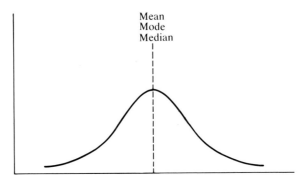

(a) Symmetrical distribution

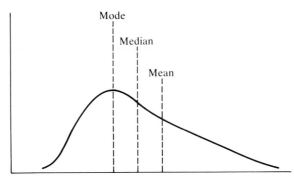

(b) Distribution skewed right

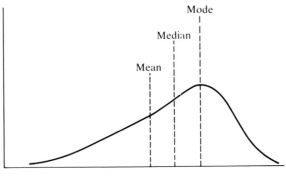

(c) Distribution skewed left

FIGURE 3.2
Location of mean, median, and mode.

Figure 3.2(a) shows a symmetrical distribution. The three values mean, median, and mode are the same and are located in the middle of the distribution. Figure 3.2(b) is skewed to the right and the mode is to the left, at the highest point of the distribution. This intuitively makes sense because the mode is the value that occurs most often. In Figure 3.2(c) the mode is located to the right of the median.

MEASURES OF DISPERSION

In the previous section we discussed how various measures of central tendency help to locate the middle of a distribution. This section discusses ways to measure the **dispersion,** or spread, of values in a distribution. Figure 3.3 shows three curves having the same mean but each having a different spread.

Curve A has less spread than curve B or C. Knowing that the mean of the three curves does not sufficiently describe the distributions, we need an indicator of the dispersion of values.

Dispersion, or spread, is important to understanding machine processes. Dispersion tells the analyst whether a machine is able to maintain a critical dimension or whether there are external causes that force the dimension to wander. Having a measure of dispersion allows a quality analyst to compare the outputs of two or more similar machines performing the same operation.

In this section we discuss the following types of dispersion: range, variance, and standard deviation.

Range. **Range** is the difference between the highest and lowest values in a data set. For example, in the lightbulb reliability array, the range is 714 hours. Mathematically, range is expressed as

$$\text{Range} = X_h - X_l$$

where X_h = highest value
     X_l = lowest value

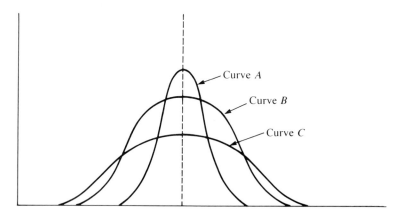

FIGURE 3.3
Measures of dispersion.

Range is easy to understand, calculate, and use. An operator monitoring a process, moving parts, inspecting parts, or adjusting the machine does not have time to calculate complicated formulas for the spread of dimensions. The range value provides a simple indicator of dispersion.

A serious drawback to the range is that it considers only two numbers and does not consider the impact of all the numbers in the data set. The range can be heavily influenced by outlying values and does not take into account how the values are clustered. For example, in a sample of observations, the analyst may make a mistake in the measurements and one outlying value can distort the calculation. This is illustrated in the following examples, which use the previous data.

EXAMPLE   Calculate the range of the following five numbers:

$$3.003 \qquad 3.002 \qquad 2.998 \qquad 2.999 \qquad 3.003$$

Solution

$$\text{Range} = X_h - X_l$$
$$= 3.003 - 2.998 = 0.005$$

If the last number in the string is changed:

$$3.003 \qquad 3.002 \qquad 2.998 \qquad 2.999 \qquad 3.048$$

the range becomes

$$\text{Range} = X_h - X_l$$
$$= 3.048 - 2.998 = 0.050$$

Variance. **Variance** is another measure of dispersion. It measures the variation from the mean of all values in a data set. Both variance and standard deviation measure the deviation, or distance, from the mean (Figure 3.4).

The population variance is the average absolute deviation calculated from a population of data. In this case we sum the squared distances between each mean and each observation and then divide by the number of elements in the population. The purpose of squaring each number is to make each number positive and not calculate the absolute value of each deviation.

What do these definitions mean? These definitions will become clearer after we have performed a calculation. The formulas for population and sample variance are slightly different. The population variance considers all observations in the population, while the sample variance considers only the observations in the sample. The observation value X is the value a quality analyst obtains by measuring a part or completing a laboratory experiment.

Population variance. The population variance, denoted by σ^2 (Greek letter **sigma**), is expressed as

$$\text{Population variance} = \sigma^2 = \frac{\Sigma(X - \mu)^2}{N}$$

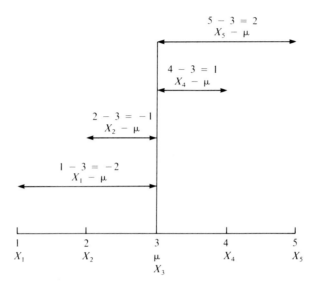

FIGURE 3.4
Deviation from population mean.

where σ^2 = population variance
Σ = summation sign
X = observation value
μ = population mean
N = total number of observations in population

This formula can be explained as follows. The expression in parentheses is the distance between any observation value and the population mean. The sum of these values is then squared and divided by the number of observations in the population.

EXAMPLE Given the following five numbers, 1, 2, 3, 4, and 5, calculate the population variance.

Solution

$$\text{Population mean} = \frac{1 + 2 + 3 + 4 + 5}{5} = 3$$

Number	$(X - \mu)$	$(X - \mu)^2$
1	$1 - 3 = -2$	4
2	$2 - 3 = -1$	1
3	$3 - 3 = 0$	0
4	$4 - 3 = 1$	1
5	$5 - 3 = 2$	4
15	0	10

$$\sigma^2 = \frac{10}{5} = 2.0$$

The calculation in the preceding example was relatively easy because we used small whole numbers. It becomes complex if numbers are large or have decimal points. Statistical calculators are available in which variance and standard deviation are programmed.

Sample variance. The sample variance formula is different from the population variance formula. The symbols are different so the analyst does not become confused. Another difference is that the denominator of the sample is $(n - 1)$ instead of N. The sample variance, denoted by s^2, is calculated as follows:

$$s^2 = \frac{\Sigma(X - \overline{X})^2}{n - 1}$$

where s^2 = sample variance
Σ = summation sign
X = observation value
n = sample number of observations
$\overline{X}$ = sample mean

Why does the denominator have $(n - 1)$ instead of N? Statistical calculations show that using n in the denominator underestimates the population variance. The expression $(n - 1)$ corrects this underestimate; the result is higher, closer to the actual population variance.

EXAMPLE   The following measurements are from the previous example. Calculate the sample variance.

Solution

$$\text{Sample mean} = \frac{3.003 + 3.002 + 2.998 + 2.999 + 3.003}{5}$$

$$= 3.001 \text{ inches}$$

Measurement	$(X - \overline{X})$	$(X - \overline{X})^2$
3.003	0.002	0.000004
3.002	0.001	0.000001
2.998	−0.003	0.000009
2.999	−0.002	0.000004
3.003	0.002	0.000004
15.005	0	0.000022

$$s^2 = \frac{0.000022}{4} = 0.0000055$$

$(5 - 1)$

EXAMPLE   These data are from the example where the last number in the string was changed. The five values are more dispersed, so logic would say that the variance (dispersion) should be higher than in the previous example.

Solution

$$\text{Sample mean} = \frac{3.003 + 3.002 + 2.998 + 2.999 + 3.048}{5}$$

$$= 3.010 \text{ inches}$$

Measurement	$(X - \bar{X})$	$(X - \bar{X})^2$
3.003	−0.007	0.000049
3.002	−0.008	0.000064
2.998	−0.012	0.000144
2.999	−0.011	0.000121
3.048	0.038	0.001444
15.050	0	0.001822

$$s^2 = \frac{0.001822}{4} = 0.0004555$$

Intuitively, this result makes sense. One value was changed from the prior example: 3.003 became 3.048. Several observations can be made. First, one outlying value, 3.048, moves the mean away from the other clustered numbers. The mean, 3.010, does not represent the clustered numbers because the 3.048 value distorts the calculation. Also, the higher number implies that the variance (dispersion) should also increase.

Standard Deviation. **Standard deviation** is derived from variance. The population and sample standard deviations are the square root of the respective variances. Population variance is expressed as σ^2 and sample variance is expressed as s^2. So population standard deviation is σ and sample standard deviation is s.

The formula for the population standard deviation is the square root of the population variance:

$$\text{Population standard deviation} = \sigma = \sqrt{\frac{\Sigma(X - \mu)^2}{N}}$$

The formula for the sample standard deviation is the square root of the sample variance:

$$\text{Sample standard deviation} = s = \sqrt{\frac{\Sigma(X - \bar{X})^2}{n - 1}}$$

When the mean and the standard deviation are computed from sampled data, the resulting numbers are called **statistics.** When the mean and the standard deviation are calculated from the population, these numbers are called **parameters.** Statistics and parameters may differ because a sample may not be representative of the population.

COMMON DISTRIBUTION SHAPES

Frequency distribution graphs can provide information that is important to quality. Specifically they indicate the central tendency of measurements, the spread of data, and the number of measurements within specification limits.

When the upper and lower specification limits are added to a frequency distribution graph, certain observations of the population of data, namely, the distribution of individual values, can be made, especially if the spread of data is compared with the specification limits. As we see in Chapter 4, this important concept is used to determine whether a process is capable of meeting specifications.

Frequency distributions have several common shapes, including symmetrical, bimodal, skewed right and left, random, and uniform (Figure 3.5).

Symmetrical. Symmetrical distributions are symmetrical about the center and have tails that trail off on both sides. If sufficient data are collected, these bell-

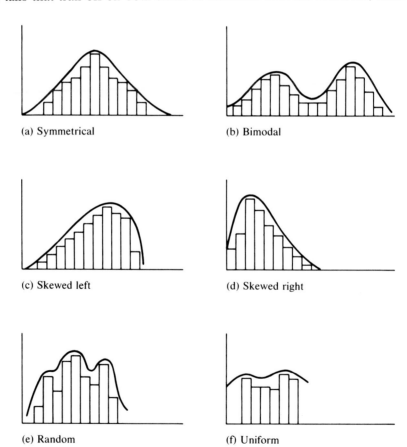

(a) Symmetrical

(b) Bimodal

(c) Skewed left

(d) Skewed right

(e) Random

(f) Uniform

FIGURE 3.5
Histogram shapes.

shaped distributions, also called normal distributions, occur frequently in business and nature [Figure 3.5(a)].

Bimodal. Bimodal distributions have two distinct humps, indicating two separate groupings of data. For example, two operators measure similar parts coming off a production line. If both sets of measurements are graphed on one chart, a bimodal distribution suggests that inspectors were using different methods of inspection or a measurement tool was out of calibration [Figure 3.5(b)].

Skewed Left. A distribution skewed left has most of its measured points on the right side of the distribution and fewer on the left. This is also called a negative skew. Possible causes of this type of distribution are: raw material has uniform high tensile strength, measuring instrument is reading high, or machine is set up improperly [Figure 3.5(c)].

Skewed Right. Similarly, a distribution skewed right has most of its measurements on the left side of the distribution. This is also called a positively skewed distribution. Causes for this type of distribution are similar to those that cause the distribution to skew to the right. The difference is that measurements gather on the left side [Figure 3.5(d)].

Random. A random distribution has no recognizable pattern. Random distributions do not reveal much information. The analyst may want to spread the data into more cells or lump data into fewer cells, and, hopefully, another recognizable distribution will emerge [Figure 3.5(e)].

Uniform. A uniform distribution has most of its measured data spread out evenly throughout the distribution. If an inspector was inspecting a batch of products 100% and rejecting those items that were outside the specification limits, the distribution of the measured data would be uniform [Figure 3.5(f)].

NORMAL DISTRIBUTION

Bell Curve

Data obtained from measurement and inspection are often distributed in a **bell-shaped** frequency **distribution.** This distribution is also called the **normal,** or **Gaussian, distribution.**

The bell curve shows us how to use the mean as a measure of central location and dispersion. Data have a tendency to cluster around a center value and to vary around that value in a known dispersion.

Bell-curve data can be analyzed graphically and mathematically. In this section we discuss the graphic elements and in the next section the analytical elements of the bell curve.

SPOTLIGHT*

When the upper and lower limits of a specification (USL and LSL) are added to a frequency distribution (histogram), important information can be derived, as illustrated in Figure 3.6.

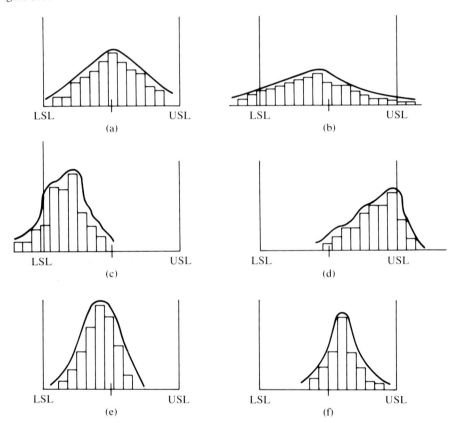

FIGURE 3.6
Histograms and specification limits.

Example (a). This histogram is evenly distributed and well centered in the middle of the specification. No defective parts are produced.

Example (b). This histogram is evenly distributed and relatively centered in the middle of the specification. However, the distribution spills beyond the upper and lower specification limits, implying that defective parts are being produced.

Example (c). This histogram is evenly distributed but is shifted to the left beyond the lower specification limit. The parts beyond the lower specification limit are defective. If the operator adjusts the process so that the histogram is centered in the middle of the specification, then all parts are acceptable.

Example (d). This histogram is shifted to the right. Again, the parts beyond the upper specification limit are defective. The same solution as in example (c) has to be used.

Example (e). This histogram is relatively well centered and is evenly distributed. The distribution spread is tightly distributed. All parts are acceptable.

Example (f). This histogram is also tight, but has shifted to the right. The operator may adjust the process and, if necessary, center the distribution. The process is showing a tendency to drift.

---

* Adapted from M. Juran and F. M. Gryna, *Quality Planning and Analysis* (McGraw-Hill, New York, 1980).

The normal curve has the following characteristics: it is bell shaped, symmetrical about the mean, and the median, mean, and mode are the same.

The normal curve represents the distribution of population data. A sample is selected from a population. It is assumed that sample data are representative, that is, show similar characteristics to the population.

If measured data are approximately distributed in a bell-shaped histogram, we can use the statistics of a normal distribution to reach conclusions about the number of products that conform to specification.

Area under the Normal Curve

Figure 3.7 shows that the percentage of the items lying between the limits of the standard deviation is known. The percentage of these items can also be calculated.

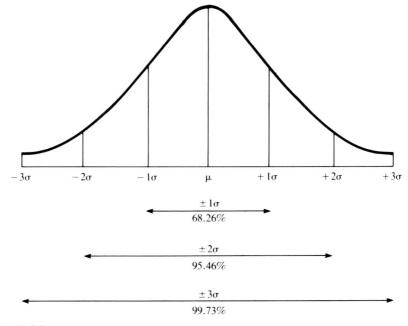

FIGURE 3.7
Standardized normal distribution.

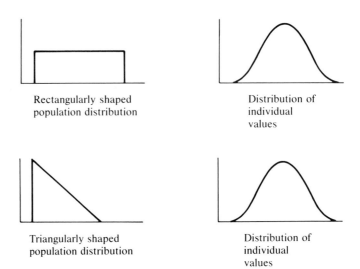

Rectangularly shaped
population distribution

Distribution of
individual
values

Triangularly shaped
population distribution

Distribution of
individual
values

FIGURE 3.8
Central limit theorem.

The important characteristics of σ and its distance from the population mean μ are as follows:

☐ 68.26% of all measurements lie between $x + 1\sigma$ and $x - 1\sigma$
☐ 95.46% of all measurements lie between $x + 2\sigma$ and $x - 2\sigma$
☐ 99.73% of all measurements lie between $x + 3\sigma$ and $x - 3\sigma$

Central Limit Theorem

Walter A. Shewhart,* a prominent statistician in the 1920s, discovered that if a distribution of measurements is from a chance or **constant-cause system,** it does not matter whether the shape of the distribution of individual values (the population) is triangular or trapezoidal. The averages of different sized samples will have a similar central tendency and will be distributed in a bell shape. This is called the **central limit theorem** (Figure 3.8). A chance or constant-cause system means that there are no abnormal factors causing measurements to vary.

Expressed in another way, the central limit theorem states that if a population is distributed in a bell shape, the distribution of a sample of the population also has a bell shape. More importantly, if the population of individual measured values is not distributed in a bell shape, the distribution of the sample still approximates a bell shape.

The central limit theorem implies that any process can be monitored and controlled over a period of time by sampling parts from a population of data and

---

* W. A. Shewhart, *Economic Control of the Quality of Manufactured Product* (Van Nostrand Reinhold, Princeton, NJ, 1931).

calculating their mean. The sample mean will be close to the population mean (Figure 3.9).

Applications

The normal curve can be used to calculate the percentage of defective products inside or outside specification limits. The areas under the normal curve are derived from the **Z values,** which are listed in Appendix A. The table can be read left to right or right to left. Areas under the normal curve are read from $-\infty$ to X_i.

First Z is calculated using the following formula:

$$Z = \frac{X_i - \mu}{\sigma}$$

where Z = standard normal value
X_i = measured value
μ = mean or average
σ = population standard deviation

Then the Z value is located in Appendix A. This represents the area under the bell-shaped curve to the left of the calculated Z value. The total area under the curve from $-\infty$ to $+\infty$ is equal to 1.0000, or 100% of the population. Appendix A lists values for only one-half of the bell curve. Since a bell curve is symmetrical about the mean, values on the left side are mirrored on the right side.

The decimal values in Appendix A have to be converted to percentages. For example, from $-\infty$ to -3σ the value is 0.135% (Appendix A shows 0.00135) of the area of the bell curve, or 0.135% of the population of items. From -3σ to the center (mean) of the curve is 49.865% of the population (0.5000 $-$ 0.00135) (Figure 3.10).

If the mean μ and the standard deviation σ of a population are known, then these techniques can be used to find the percentage of items that are either less or greater than a value or that are between two values.

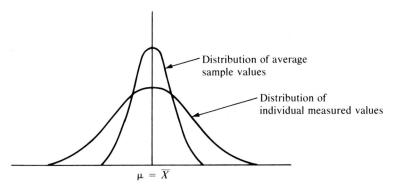

FIGURE 3.9
Distributions of individual values and sample averages.

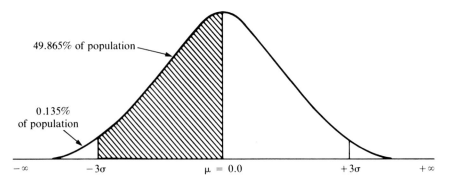

FIGURE 3.10
Example of Z values.

Less Than a Value. This technique is demonstrated using the following data. The mean weight μ of a shipment (population) of raw steel bars is 10.1 pounds. The standard deviation σ of the shipment is 0.09 pound. The calculation is illustrated in Figure 3.11. The first step is to determine the Z value using the Z-value equation. The variables are known, so values are simply substituted into the equation to provide the Z value. From Appendix A it is found that $Z = -3.333$ equals 0.00043, or 0.043%. Thus 0.043% of the steel bars fall below 9.8 pounds.

Greater Than a Value. Using the same data and following the same process, the percentage of items greater than 10.2 pounds is found to be 13.4% (Figure 3.12).

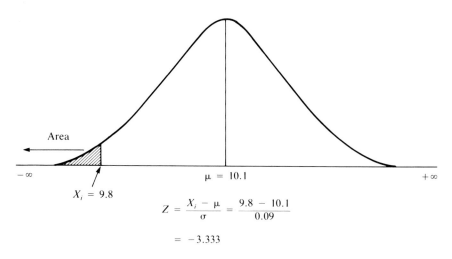

FIGURE 3.11
Example problem: less than a value.

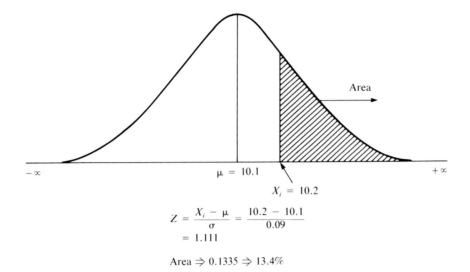

FIGURE 3.12
Example problem: greater than a value.

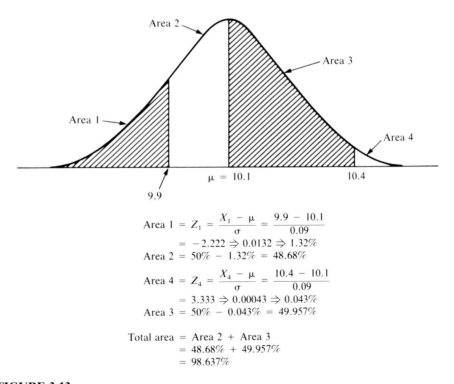

FIGURE 3.13
Example problem: between two values.

Between Two Values. To determine the percentage of items that fall between two given values, the quality analyst calculates first the percentage of items less than the lower of the two values and then the percentage of items greater than the higher of the two values to find the percentage that lies between these two values. Referring to Figure 3.13, the bell curve is divided into four regions, two on either side of the centerline. The analyst first calculates area 1 using the "less than a value" technique discussed. Area 2 is determined by subtracting this value from 50%, which is half of the bell curve area. Next the analyst calculates area 4 using the "greater than a value" technique and subtracts this value from 50%, which provides area 3. Finally, the analyst adds areas 2 and 3 to find the total area between the two values. The percentage of items between 10.4 and 9.9 pounds is 98.637% (Figure 3.13).

SPOTLIGHT

In a competitive global economy, organizations are attempting to satisfy the customer with quality products. Ford, as anyone can see in the television commercial, believes that "quality is job 1." This sometimes seems like commercial hype. However, the operating profits of Ford over the last several years attest to the quality emphasis that Ford is promoting. This is illustrated in the following quotation:

> The operating philosophy of Ford is to meet customer needs and expectations by establishing and maintaining an environment which encourages all employees to pursue never ending improvement in the quality and productivity of products and services throughout the company, its supply base, and its dealer organizations.*

An important element of Ford's quality program is the pursuit of quality with its supplier base. A large portion of Ford's manufacturing dollar goes to outside suppliers. If a company wants to produce a quality product, the material supplied must conform to specifications.

Ford, GM, and many large companies use a weighted mean method for evaluating suppliers. When a product is purchased, suppliers are evaluated by criteria that commonly include quality, price, and service.

Before a supplier is evaluated, representatives of engineering, manufacturing, and purchasing weigh the criteria in terms of relative importance. A supplied product that must perform reliably in different environments will have its quality criterion weighed more heavily than other criteria.

In our example, the representatives weighed the criteria as follows:

☐ Quality = 40 points
☐ Price = 30 points
☐ Service = 30 points

Three vendors are evaluated after a year of supplying the same product. First the quality factor is determined in terms of material shipments that were accepted or rejected based on incoming inspection. The results are:

---

* Ford Motor Company, *Quality System Standard*, Q-101, p. 4, 1984.

	Lots Received	Lots Accepted	Lots Rejected	Percentage Accepted	× Quality Factor	= Quality Rating
Supplier A	40	36	4	90	0.40	36
Supplier B	30	24	6	80	0.40	32
Supplier C	25	25	0	100	0.40	40

Next price is considered. Price is normalized to reflect the importance of the lowest price. This means that the lowest price has the highest rating and the other prices are rated against the lowest price. Otherwise the highest price in the calculation would receive the highest value.

	Lowest Price	Net Price	Percentage	× Price Factor	= Price Rating
Supplier A	$2.00	$2.00	100	0.30	30
Supplier B	$2.00	$2.20	91	0.30	27.3
Supplier C	$2.00	$2.10	95	0.30	28.6

The third area is service. Service quality is based on promises kept. The service factor is 0.30.

	Promises Kept	× Service Factor	= Service Rating
Supplier A	90%	0.30	27
Supplier B	100%	0.30	30
Supplier C	100%	0.30	30

Finally, a composite rating of quality, price, and service is developed. The composite rating weighs the totals of the three factors, and the highest score gets the following year's contract.

Rating	Supplier A	Supplier B	Supplier C
Quality (40)	36	32	40
Price (30)	30	27.3	28.6
Service (30)	27	30	30
Total rating	93	89.3	98.6

Supplier C with the highest quality and service but also the second highest price is awarded the contract. It should be noted that supplier C's product cost is competitive with that of the other two suppliers. If supplier C's product had been priced exorbitantly, this would have been reflected in a lower score, which could ultimately have given another supplier the contract.†

---

† For additional discussion see *Procurement Quality Control*, 3rd Edition, Milwaukee: American Society for Quality Control, 1985 pp. 117–133.

SPOTLIGHT

Just-in-time (JIT) manufacturing is closely linked to total quality improvement. Simply, JIT is the delivery of material to the right location just in time to be used.

As simple as it sounds, there are many implications to this statement. The supplier has to send defect-free material to the right loading dock, properly packaged, within narrow time windows. If the material is defective, the shipment is rejected and sent back to the supplier. If the plant does not have any of the required material in inventory, the production line shuts down.

Stockless Production

JIT is sometimes called stockless production or zero inventory because there is no storage of material. However, these JIT definitions are too narrow.

As you can see, JIT is more than an inventory-reduction technique. JIT has evolved into an organizational philosophy, stressing excellence in engineering, production, quality assurance, purchasing, and materials management.

Waste Elimination

In Chapter 1 we discussed the prevention and the inspection modes of quality control. In the prevention mode, a process is controlled so that defects are prevented from occurring. In the inspection mode, products are accepted or rejected based on the results of inspection. Rejection adds unwanted costs in terms of scrap, rework, or simply waste.

An important concept in JIT is the elimination of all types of waste. Waste includes all activities that add cost or reduce product or service value. In particular, JIT stresses the elimination of the following types of waste: defects, late or early deliveries, excessive machine set-up time, redundant handling, high inventories, and long lead times. The following discussion illustrates some of the areas that JIT impacts.

Zero-defective material is required from suppliers. A shipment of defective material can stop a production line. There is no material in inventory and production will restart when a shipment of acceptable material arrives. The need for defect-free material requires a close working relationship between supplier and customer.

All inventory, including raw material, buffer, and final product, is considered waste. By eliminating all types of inventory, problems are discovered that were hidden previously. Inventory may hide poor-quality material, late delivery of parts, machine scrap, and other problems.

As an illustration, inventory often hides poor-quality material and creates an unacceptable recurring situation. Any time there is a shipment of nonconforming material, it has to be segregated and moved into inventory; the supplier is notified of the problem; paperwork must be generated; nonconforming material is moved onto the shipping dock and sent back to the supplier; and the next shipment will probably be sampled and inspected. All of this activity can be eliminated if the supplier provides zero-defect material.

A shipment must also be delivered within a narrow time window. If the shipment is delivered late, the production line stands idle and money is lost. If the shipment arrives too early, the driver may wait because there is no room and nobody is available to unload the truck.

The shipment must also be properly packaged. If a shipment is packaged loosely, material may be damaged in transit. If the packaging is too secure, time (money) may be lost unpacking and getting the product onto the production line.

Inside the factory, JIT includes all activities to ensure that products are delivered to the next work station just in time to be processed. There is little or no waiting between manufacturing steps. Unnecessary material movement and storage add cost. All restrictions to the smooth flow of material are eliminated.

SUMMARY

- Central tendency and variation are two methods for describing and measuring numerical data. By using measures of central tendency it is possible to describe the center of a set of observations or measurements. Variation and dispersion are similar terms describing the spread of data.
- Three common measures of central tendency are mean, median, and mode. Mean is the average of a set of values. Median is the center of an array of numbers. Mode is the value that appears most often in a set of numbers.
- Three common measures of variation are range, variance, and standard deviation. Range is the difference between the highest and lowest observed values. Variance is the average of the squared distance between the mean and each observed item in a population. Standard deviation is the square root of the variance.
- Standard deviation is the preferred method for calculating variation. It assesses the spread of data more accurately by lessening the influence of large or small numbers. Standard deviation is difficult to compute without a calculator.
- Range is another method for calculating variation. Range is easy to use and to understand.

- If measurements from a process or machine are graphed, they often form a bell-shaped frequency distribution. This distribution is also called a normal, or Gaussian, distribution.
- A normal curve has three important characteristics. It is shaped like a bell. The curve is symmetrical about the mean. The median, mean, and mode are the same.
- Variation of the normal curve is measured by the standard deviation. Sigma (σ) is the unit that measures the standard deviation. Three important characteristics of sigma are: 68.3% of all observations lie between $\pm 1\sigma$, 95.5% of all observations lie between $\pm 2\sigma$, and 99.7% of all observations lie between $\pm 3\sigma$.
- The central limit theorem states that if all measurements are from a constant-cause system, the average value from a sample pulled from different-shaped distributions has the same central tendency as the population and is spread in a bell-shaped distribution.
- Using the bell curve, it is possible to calculate the percentage of the population that falls outside of specified limits.

KEY TERMS

bell-shaped distribution Same as normal distribution.

bimodal distribution Curve with two modes.

central limit theorem Theorem stating that, regardless of the shape of a distribution of individual values, a sample taken from this distribution will have a similar central tendency and will be distributed in a bell shape.

common-cause system Process in which variations are random, occur by chance, and are constant over time. See constant-cause system.

constant-cause system Process in which variations are random, occur by chance, and are constant over time.

dispersion Spread, or variation, of data.

Gaussian distribution Same as normal distribution.

mean Measure of central tendency; calculated by adding all of the values in a set and dividing by their number; also called average.

measure of central tendency Tendency of measured or observed data to cluster around a central value.

median Measure of central tendency; value that falls in the middle when values are arranged in an array.

mode Measure of central tendency; value that is repeated most often in a data set.

normal distribution Tendency for a large number of observed data in business, quality, and nature to be distributed in a bell shape; also called Gaussian or bell curve.

parameters Measures that describe the characteristics of a population.

range Measure of dispersion; the highest value minus the lowest value in a data set.

sigma (σ) Unit of measure of dispersion of data.

standard deviation Measure of dispersion; square root of variance.

statistics Measures that describe the characteristics of a sample.

symmetrical distribution Example of bell curve.

variance Measure of dispersion; average squared distance between the mean and each value in the population.

weighted mean Mathematical calculation weighing relative values in terms of their importance.

Z values Values used to estimate area under bell curve.

QUESTIONS

1. Is a sample pulled from a population of products necessarily representative of the population? Explain.

2. What are the common measures of central tendency? Explain each.

3. In which calculation, mean, mode, or median, is every observation value taken into consideration?

4. Do extreme values have an effect on the calculation of mean, mode, or median? Explain.

5. If a distribution is skewed to the left, where are the mean, mode, and median located?

6. What are the advantages and disadvantages of using the mean, median, and mode as measures of central tendency?

7. When should the weighted mean be used in place of the simple mean?

8. How do you find the median of an array with an even number of elements? With an odd number of elements?

9. What is a bimodal distribution? Draw one.

10. What are common measures of variation? Explain each.

11. What are the comparative advantages and disadvantages of using different measures of variation?

12. How is the standard deviation calculated?

13. What are characteristics of the normal curve? Explain each.

14. What is σ and how does it relate to distances on the normal curve?

15. If a population of items, when graphed, is shaped like a square, how would the sample from this population be distributed?

PROBLEMS

1. In each of the following samples, calculate the mean, median, and mode (modes):
   a. 2, 4, 5, 7, 8, 10, 15
   b. 2, 4, 6, 7, 11, 14, 19, 20
   c. 1, 2, 2, 3, 5, 6, 8, 8, 9
   d. 1, 2, 3, 4, 5

2. In each of the following samples, calculate the mean, median, and mode (modes):
   a. 1, 3, 5, 6, 8
   b. 0, 3, 5, 7, 9, 10
   c. 1, 2, 3, 4, 4, 5, 6, 6, 7
   d. 2, 3, 5, 7, 8, 8, 9

3. In the following sample, calculate the range, variance, and standard deviation:
   2, 4, 6, 9, 11, 12

4. In the following sample, calculate the range, variance, and standard deviation:

   1, 2, 3, 4, 5, 6

5. In the following sample, calculate the range, variance, and standard deviation:

   12, 16, 16, 24, 26, 28, 30

6. In the following sample, calculate the range, variance, and standard deviation:

   15, 18, 19, 20, 22, 28, 30, 32

7. A sample of products pulled from an incoming shipment of material has the following weights:

   3.01, 3.05, 2.96, 2.95, 2.99, 3.02, 3.00

   Find:
   a. Mean
   b. Median
   c. Mode
   d. Range
   e. Variance
   f. Standard deviation

8. Steel samples were tensile tested. The following are results of the test; units are in thousands of pounds per square inch:

   60, 61, 58, 59, 61, 62, 58, 57, 59

   Find:
   a. Mean
   b. Median
   c. Mode
   d. Range
   e. Variance
   f. Standard deviation

9. A shipment of products consisted of six lots of differing sizes. The shipments were 100% inspected for conformance to specifications. The percentages of defective items in the seven lots are given below. Find the percentage of defective items of the whole shipment.

Lot Size	Percent Defective
50	0
100	15
200	6
50	8
150	5
100	4
50	2

10. The population mean is 100 pounds for aluminum ingots. The population standard deviation is 10 pounds. The distribution is normal. Determine the percentage of the population less than 94 pounds, greater than 120 pounds, and between 80 and 120 pounds.

11. The mean length of a lead pencil is 8 inches. The distribution is normal. The population standard deviation is 0.015 inch. Determine the percentage of the population of pencils

that is less than 7.96 inches, greater than 8.05 inches, and between 8.05 and 7.96 inches.

12. A shift supervisor pulled five parts from a production lot with the following measurements: 2.002, 2.001, 1.999, 2.003, 1.998. The specification limits are 2.000 ± 0.005 inches. What are the sample mean and the sample standard deviation of these numbers? Based on this example, is the process manufacturing parts to specification?

13. In the next hour, a shift supervisor pulls five more products from production with the following measurements: 2.002, 2.005, 2.005, 2.003, 2.004. The specification limits are 2.000 ± 0.005 inches. What are the sample mean and the sample standard deviation of these numbers? Based on this example, is the process manufacturing parts to specification? What is happening to the process?

14. A machining operation is producing quarter-inch bolts with an average outside diameter of 0.251 inch and a population standard deviation of 0.002 inch. If the distribution is approximately normal, what percentage of the production is falling outside of the specification spread of 0.250 ± 0.004 inch?

15. Referring to Problem 14, the process average was centered at 0.250 and the population standard deviation was tightened to 0.0016 inch. If the distribution is approximately normal, what percentage of the production is falling outside of the specification spread of 0.250 ± 0.004 inch?

16. Using the data of Problem 15, engineering tightened the specification to 0.250 ± 0.003 inch. What percentage of the production is falling outside of the specification spread?

17. Three grades of products, their volumes, and the prices are shown. Calculate the weighted mean cost of the product.

Type of Product	Cost of Material	Number Sold
Good	$20.00	800
Better	$24.00	100
Best	$30.00	100

4

Statistical Process Control: Variable Charts

The statistical concepts we have covered lead us to statistical process control (SPC). SPC uses a measure of central tendency and a measure of dispersion to monitor **processes** instead of inspecting results after a process has produced a product. Specifically, **average** tracks the central tendency and **range** tracks the dispersion of sampled measured data in a quality characteristic.

In Chapter 1 we discussed the relative advantages of prevention contrasted to inspection. Defect prevention is the goal of company-wide quality management programs. Inspection is being eliminated because of the high cost of scrap and rework.

SPC is a major prevention technique by which a process is monitored and controlled instead of being inspected after the fact to catch defects. SPC can be used to make decisions about a machine, chemical, electronic, or thermal process.

OVERVIEW OF STATISTICAL PROCESS CONTROL

SPC encompasses the following basic ideas:

- ☐ Quality is conformance to specifications.
- ☐ Processes and products vary.
- ☐ Variation in processes and products can be measured.
- ☐ Variation follows identifiable patterns.

☐ Variation due to assignable causes distorts the bell shape.

☐ Variation is detected and controlled through statistical process control (SPC).*

Quality is Conformance to Specifications. In Chapter 1, we defined quality as conformance to specifications, customer satisfaction, and value added. Using the conformance to specification definition, quality is assured by measuring a product's critical dimensions or a process's output and comparing these against a specification or engineering tolerance. The conformance definition eventually must be defined in terms that can be understood by the people who produce the product or deliver the service.

Engineering develops a set of specifications for the product, assemblies, subassemblies, systems, subsystems, and components. Specifications at the product level detail acceptable levels of performance, reliability, maintainability, serviceability, appearance, ease of use, or safety. At the constituent component level, specifications detail acceptable dimensions, fit, and interaction of components. Specifications for a complex product consist of thousands of pages.

Products and Processes Vary. It has been said that the control and assurance of quality are the study of variation, the dispersion of measurements in a product quality characteristic. Measured data have a tendency to cluster around a point and be dispersed around this point. These tendencies are found in business and in nature. In manufacturing, no two machines are the same. Even no two parts made from the same machine are identical. There is always some variation that may only be detected at a microscopic level.

Variation Can Be Measured. Variation can be detected and measured by critical measuring instruments. Some instruments can measure to millionths of an inch. These instruments must be stored, handled, and used carefully. Since dust, heat, or vibration can affect an instrument's precision, special "clean rooms" are used, which are humidity, particulate, and temperature controlled environments.

Variation Follows Identifiable Patterns. Variation follows identifiable patterns which often are in a bell-shaped frequency distribution. The bell curve is the distribution one would expect if there were no external assignable factors influencing the process. This means that all measurements follow a random pattern.

Variation Distorts the Bell Shape. Variation due to assignable causes distorts the bell curve. The distortion looks like shapes that were shown in Chapter 3. For example, material consistently hard or measuring instruments reading high would indicate that the distribution is skewed to the left.

Variation is Detected and Controlled Through SPC. Variation can be detected and controlled through SPC. This method, based on statistical principles, allows

---

* Derived from G. Hutchins. SPC(QPE, Portland, 1989).

an operator to detect deviations in a measured value and then adjust the process back to the target dimension.

TYPES OF DATA

Attribute Data

Data can be classified by how they are collected. Data obtained from measuring instruments that either accept or reject a part are called **attribute data.** Attribute data are classified in terms of good/bad, acceptable/unacceptable, conforming/defective, or go/no-go. Attribute charts, the subject of Chapter 6, are used to track attribute data.

Attribute data are often collected in the final inspection, when assembled products are performance tested or inspected visually. Attribute data do not provide detailed information for improving a process. They are mainly used to summarize the results of an operation. If problems arise, attribute data indicate that something is amiss in a machine, process, or work area. For example, attribute charts track the number of defects or the percentage of defectives. The data do not reveal the amount of deviation from **specification limits,** also called **tolerance limits.**

Variable Data

Data derived from incremental measurements are called **variable data.** Quality improvement presupposes the ability to distinguish incremental differences in dimensions of a quality characteristic. This requires the collection of variable data. **Variable control charts** illustrate variable data. The following example should delineate the differences between variable and attribute data.

Go/No-Go Measuring Instruments. Attribute data will indicate whether a part conforms to a specification or is nonconforming. But variable data will reveal to what extent a measured dimension conforms to the specification.

The following example illustrates the difference between the measurement of variable and attribute data. Go/no-go gages are measuring instruments used to measure attribute data. They indicate only two conditions, whether a part is acceptable or unacceptable.

A round bar's outside diameter (O.D.) specification is 3.000 ± 0.010 inches. If a **micrometer** is used to measure the outside diameter of the bar, possible measurements could be 3.000, 3.009, 2.998, or 2.987 inches. These are all variable data or variable measurements.

Instead of using a micrometer to measure the part, two round gages are designed to fit over the round bar. One gage has an opening slightly larger than the upper specification limit; for example, it is 3.0101 inches wide. Another gage has an opening slightly smaller than the lower specification limit, 2.9899 inches wide.

The inspector uses both gages to evaluate the conformance of the part. The inspector first fits the larger gage over the round bar. If the gage fits over the bar, the bar conforms to the upper specification limit. If the gage interferes with the bar, the outside diameter of the bar is greater than the upper specification limit. Next the inspector fits the smaller gage over the round bar. If the smaller gage fits over the bar, the bar is below the lower specification limit. If the smaller gage interferes with the bar, the outside diameter of the bar is greater than the lower specification limit.

Go/no-go measurement is simple and fast. The inspector does not have to know how to use a micrometer. The two measurements can be done quickly. The problem with attribute data measurement is that gages do not tell the inspector where the dimensions are centered and how far dimensions vary inside or outside specification limits.

CAUSES OF VARIATION

In the 1930s Shewhart,* an eminent American statistician, developed the Shewhart control chart, later abbreviated to control chart. The **control chart** documents the average location and the dispersion of measured data. The control chart indicates whether a process is "in control" or "out of control," which is explained later in the chapter. The control chart also signals when a process has shifted from a specified level and corrective action is necessary. Shewhart identified two types of causes of variation, chance (random) causes and assignable (special) causes.

Chance Causes

Chance causes, also called **common causes,** are a natural part of the system. All service and manufacturing processes have natural variations. A telephone operator may answer the phone in a similar manner each time. The operator may be eating a sandwich, but her method of answering the phone changes little. A machine operator produces parts in a definite manner by following a prescribed procedure. Both operators perform their tasks without any change until some external cause forces a change. If the telephone operator had a cold or the machine operator was supplied with hard raw material, both workers' output would vary.

Assignable Causes

Once a pattern of behavior or a pattern of production is established, any abnormal deviation from the pattern is due to an **assignable cause.** If a customer on the phone complains or if defective products are produced, the telephone operator or the machinist will change their behavior patterns to eliminate the deficiencies.

Assignable causes are not random. They produce erratic behavior for which a cause can be identified. The nonrandom pattern can be seen on a control chart.

---

* W. A. Shewhart, *op. cit.*

If the pattern is entirely due to chance causes, it shows up as a bell-shaped histogram. Assignable causes would distort the bell shape. These causes disturb the system to create a nonsymmetrical distribution. Once these unnatural causes are identified, they can be eliminated.

CONTROL CHART FUNDAMENTALS

Variable Control Charts

This section outlines the mechanics of drawing a variable control chart and then explains the basic elements of the variable control chart. The variable control chart, also called $\overline{X}$ and R chart, is a graphic method of monitoring changes in central tendency and variation (dispersion) of a set of sampled product quality measurements. The variable control chart consists of two graphs. The $\overline{X}$ **bar chart** graphs the mean, or average, values of a sample of measurements over time. The **R chart** graphs the variation of the measurements in each sample over time. Most $\overline{X}$ and R charts have the $\overline{X}$ graph above the R graph on the same sheet of paper (Figure 4.1).

$\overline{X}$ and R charts can be used to:

- ☐ Monitor and control machines and processes.
- ☐ Obtain information about specifications and manufacturability. Charts graphically display a relationship between measured data and specifications.
- ☐ Obtain data about a production run. If the process operator is using control charts to maintain process control, the chart provides a level of assurance that quality is being maintained.
- ☐ Supply information to customers of conformance to specifications. As firms eliminate incoming material inspection, they demand proof of conformance by submitting control charts of the specific quality characteristic.
- ☐ Provide a visual baseline for continuing process improvement over time.

Introduction to Control Charts

The mechanics of constructing a control chart can be illustrated by a machining operation of the outside diameter of a rotor pin. At 7:00 A.M. the operator (machinist) sets the machine at the target (nominal) value of the specification. The target value is 50.0 millimeters (mm), which is the middle of the specification spread. The specification for the quality characteristic is 50.0 ± 20.0 mm. The production run is 2500 parts at 100 parts per hour. The operator takes 25 sets of measurements.

After the first part, the operator measures the product dimension being machined to ensure that dimensions are at the target value. If there has been some abnormal or unnatural drift, the operator adjusts the process back to the target.

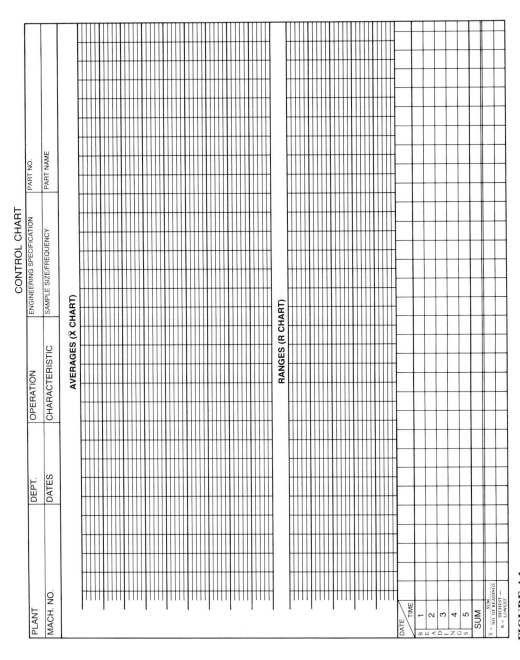

FIGURE 4.1
Blank $\bar{X}$ and R chart.

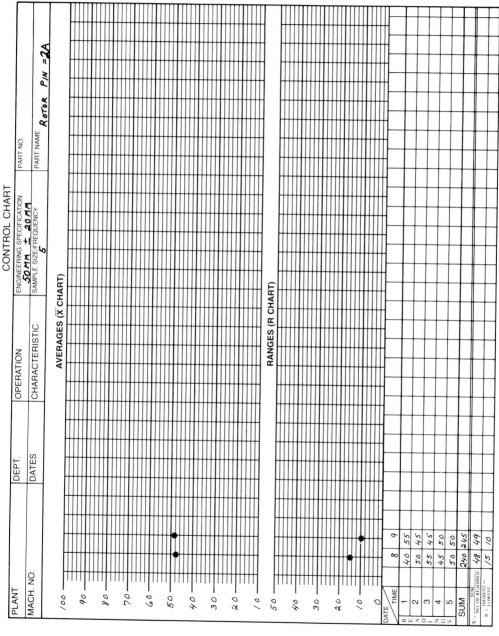

CONTROL CHART

| PLANT | DEPT. | OPERATION | PART NO. |
| MACH. NO. | DATES | CHARACTERISTIC | |

ENGINEERING SPECIFICATION
$30 MM \pm 20 MM$

SAMPLE SIZE/FREQUENCY
5

PART NAME ROTOR PIN = 2A

AVERAGES (X̄ CHART)

100
90
80
70
60
50
40
30
20
10

RANGES (R CHART)

50
40
30
20
10
0

DATE			
TIME		8	9
R E A D I N G S	1	40	55
	2	50	45
	3	55	45
	4	45	50
	5	50	50
SUM		240	245
X̄ = SUM / NO. OF READINGS		48	49
R = HIGHEST – LOWEST		15	10

FIGURE 4.2
X̄ and R chart example.

77

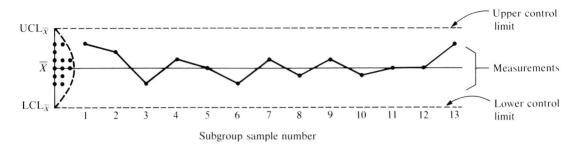

FIGURE 4.3
Vertical histogram.

Then every hour or every 100 parts, the operator pulls five more parts and measures the outside diameters. The five parts are called a **subgroup sample.**

At 8:00 A.M. the operator pulls a sample of five products with the following measurements: 40, 50, 55, 45, and 50 mm. The operator calculates the sum, range, and average of these measurements and records the data on the chart. The subgroup sample sum is 240, the subgroup sample average is 48, and the subgroup sample range is 15 (Figure 4.2).

Histogram. When the data points plotted on the $\overline{X}$ and R charts are accumulated along the left sides of the graphs, a histogram develops similar to those discussed in Chapter 3. This histogram is on the vertical axis instead of the familiar horizontal axis (Figure 4.3).

The process may be in control or **out of control.** A process in control is characterized by plotted points being randomly distributed inside the **control limits.** Most of its points are in a bell-shaped histogram inside the control limits. If the process is out of control, the histogram will be skewed with nonrandom points or with points outside the control limits (Figure 4.4).

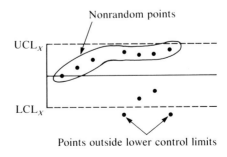

FIGURE 4.4
Skewed histogram.

Visit to a Control Chart

If the $\overline{X}$ chart is redrawn following a production run, the **centerline** is the mean, or average, calculated from previous production data. If the $\overline{X}$ chart is being drawn for the first time, the operator collects 20 to 25 samples, each of which consists of four to five measurements, and calculates the trial process average and control limits for the $\overline{X}$ and R charts. If the process is out of control, **the trial control limits** will change until the process is in control. The process average can be thought of as the mean, or average, of all the averages. The process average on the $\overline{X}$ chart is $\overline{\overline{X}}$, on the R chart it is $\overline{R}$.

The two dashed lines on each chart are the **upper control limit** (UCL) and the **lower control limit** (LCL). On the $\overline{X}$ chart, the limits are designated UCL_X and LCL_X, on the R chart, they are UCL_R and LCL_R. These limits are calculated from samples taken from previous production runs.

The normal distribution has an average, or mean, that is the process average; it is shown as the solid centerline. The upper control limit at $+3\sigma$ and the lower control limit at -3σ are calculated from the subgroup sample averages and are similar to the bell-curve σ limits.

In Control or out of Control?

At this point the process may appear to be either in statistical control or out of control. A process in statistical control has its measurements plotted inside the calculated control limits. If plotted points are inside the control limits and follow a random pattern, variation is natural. If the points follow a nonrandom pattern inside the limits or are outside the limits, an assignable cause is present.

As long as parts are acceptable, the operator leaves the process alone. If the process has been in control for a period of time, the operator can reduce the frequency of inspection. This decision is usually discussed with the area supervisor or the process engineer.

If the operator recognizes a nonrandom pattern of measurements, or a point outside the control limits, the operator identifies the cause, adjusts the process, and eliminates the root cause. Usually identifying the immediate cause and correcting the process can be done during production. Eliminating root causes requires more time.

In the previous chapter we illustrated this point with the example of a failed pump. The immediate cause of the failure was a faulty pump seal. The root cause was an inadequate preventive maintenance program.

Specifications and Control Limits

Sometimes specification limits and control limits are confused and thought to be similar. They are different and independent of each other. Specifications are detailed in drawings, documents, or regulations. They prescribe what is acceptable for a product characteristic in an individual component or product. For example, a specification may describe dimensional tolerances of a gear, and another specification may describe the performance of the interaction of many gears in a transmission.

On the other hand, control limits are calculated from measurements of products being machined or processed. Control limits are based on the dispersion in the measurements in each subgroup sample, not in measurements of individual products.

SPC PLANNING

Control charts are often misused. Control charts are installed as a quick fix to improve quality. A manager goes to a statistical process control seminar and hears success stories. Or a customer reduces incoming material inspection and demands documentation of process control. Unfortunately, unless **quality planning** is practiced, results and the subsequent analysis may be flawed.

In order to develop a reliable SPC system, the following steps should be followed before SPC is implemented:

  □ Obtain management support.
  □ Define production or service system.
  □ Identify specifications and standards.
  □ Identify product characteristics.
  □ Identify locations causing variation.
  □ Define measurement system.
  □ Focus on continuous improvement.*

Obtain Management Support. Top management must support the idea of statistical process control. Process controls require internal self-discipline, training of personnel, and new attitudes from supervision and operators.

This sometimes is difficult for first-line supervision, quality control inspectors, and machine operators. If these groups resist and view SPC as an unnecessary intrusion, it will not work. First-line supervisors may see SPC as an implied criticism of their operational management. Quality control inspectors may view this as job elimination. Process or machine operators may see it as unnecessary work.

These problems can be prevented if top management explains the purpose and importance of SPC in achieving continuous improvement. Furthermore, SPC may mean additional duties and retraining for some workers and supervision. It is essential that the union be a partner in the process of this change. If union and labor perceive this as a threat, the SPC program will fail.

Define Production or Service System. A production system is a series of step-by-step operations consisting of machine, inspection, fabrication, and assembly operations. The system may be only 1 step or 100 steps. A service system is a

---

* G. Hutchins. Op. cit.

series of steps ranging from training to dealing with the customer to delivering a product. In both systems each step has to be defined in terms of what is the operation, who does what for whom, when is it done, where does the product go next, and why is the process operating the way it is?

Identify Specifications and Standards. In terms of a manufacturing perspective, quality is conformance to specifications. Quality characteristics are specified in an engineering drawing, specification, industry standard, or customer drawing.

If a product is acceptable, its measured dimensions are inside the tolerance spread. Often specifications call out a target, or nominal, dimension and upper and lower specification limits. The upper specification limit minus the lower specification limit is the tolerance spread, or simply the tolerance. A specification can detail surface finish, workmanship, dimensions, performance, reliability, or maintainability.

Identify Product Characteristics. A product has many characteristics. A variable control chart measures only one product quality characteristic. Multiple charts tracking multiple quality characteristics can be developed. It is more useful to monitor one significant characteristic per operation, such as a physical dimension, weight, strength, reliability, or durability.

Identify Locations Causing Variation. Processes and products vary. Manufacturing processes are composed of many variables, which are the elements of the cause-and-effect (C–E) diagram: machines, methods, operator, materials, and environment.

Before production, the dispersion of values from a quality characteristic should be understood and minimized if a process is to produce products conforming to specifications. The goal is to create uniformity in each variable so that any abnormal variation can be discovered. This means that machines must be capable of manufacturing products. Accurate procedures and work instructions are provided to the workers. Operators are thoroughly trained on the machines. Incoming raw material is uniform and conforms to specifications.

Define Measurement System. The measurement system consists of elements that ensure accurate and precise measurement, including identifying the appropriate unit of measurement, obtaining proper test and measurement equipment, and training personnel. A product characteristic may be measured to ten thousandths or even millionths of an inch.

Test and measurement equipment is chosen with the required precision and accuracy. A special testing facility, such as a clean room, may be built to accomplish the testing. The room may be dust, temperature, and humidity controlled. Personnel may also be trained in the use, handling, and storage of the equipment.

Focus on Continuous Improvement. Sometimes control charts are installed to track an existing process. As long as products are made to specification and few defective products are produced, the chart is maintained and the process is not touched. This is a static approach to quality control. The goal of process charts is first to control the process, and then to improve it over time. Eventually manufacturers strive to attain rates of defects per 10,000 or one million products. Process limits are becoming progressively more narrow over time so that most subgroup sample averages eventually are grouped about the specification target.

CONSTRUCTION OF A TRIAL CHART

With proper planning and a rudimentary knowledge of statistics, simple $\overline{X}$ and R charts can be drawn. Initially a trial chart is constructed because it is not known whether the process is in control or capable of meeting specifications.

In this section the construction of a trial chart is explained by means of Figure 4.5, using the earlier example of a rotor pin. This process is being run for the first time. There is no production history.

The following steps for constructing the trial $\overline{X}$ and R control charts are explained in this section:

- ☐ Obtain data.
- ☐ Choose subgroup sample size and frequency of measurement.
- ☐ Calculate average $\overline{X}$ and range for each subgroup sample.
- ☐ Select scales for $\overline{X}$ and R charts.
- ☐ Plot values.
- ☐ Calculate average range and process average.
- ☐ Calculate trial control limits.
- ☐ Construct chart.
- ☐ Interpret trial limits.*

Obtain Data. Once a blank $\overline{X}$ and R chart is obtained, the operator fills out the top boxes. Since the SPC chart documents the production history of a product's quality characteristic, background information on the process is essential. Part name, operator's name, machine name, part number, drawing number, and specification limits may be written on a chart.

This background information is required for several reasons. SPC charts are sent to customers to attest that process controls were implemented. They are used as legal documentation that a product met specifications. Also, if a product failure occurs in the future, the charts indicate who did what, when, and where. A standard blank $\overline{X}$ and R chart is given in Figure 4.1.

As described, the operator sets the process or machine to the target, or nominal, dimension. The first item from the process is checked to ensure that product dimensions are at the nominal value. Then, periodically, samples are pulled from production and tested. Data can be obtained from inspection, test, or measurement.

---

* G. Hutchins. Op. Cit.

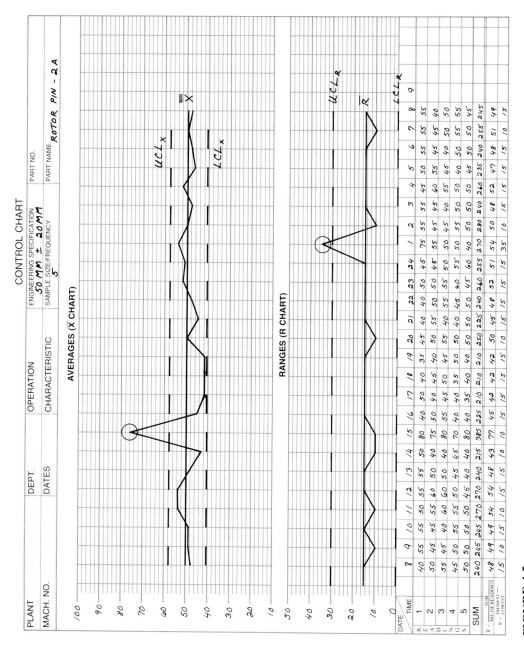

FIGURE 4.5
Completed $\bar{X}$ and R chart.

These data are used to construct the $\overline{X}$ and R charts. The $\overline{X}$ chart graphs the averages of the measurements of each subgroup sample. The R chart graphs the ranges of each subgroup sample.

The specification for the rotor pin is 50 mm ± 20 mm. This is a broad specification spread, useful for illustrating the mechanics of constructing a chart. Realistically, specification limits for a similar rotor would be 50 ± 0.1 mm. However, such calculations would be complicated and detract from the explanation of constructing these charts.

Choose Subgroup Sample Size and Frequency of Measurement. The subgroup **sample size** can vary from three to ten products. Determining the sample size is a decision made by a process engineer or the area supervisor. The decision is based on production rate, cost of sampling, ability to perform measurements, and ability to detect process changes.

The frequency of sampling can be every 10 minutes or every hour. The number of subgroups should be large enough to reveal changes in the central tendency and variation of data between subgroups. In other words, sufficient subgroups should be tested so that assignable causes of variation have a chance to appear. As a practical matter, 25 or 30 sample subgroups should be selected from a production, each having three to five measurements.

Initially subgroup samples are selected in frequent intervals to establish a process baseline. The operator does not know whether the process is in statistical control or whether it can meet specifications. If the process is in control and has shown stability over time, sampling intervals can be less frequent, and the sample size can be smaller.

In our example the process engineer determined that the sample subgroup size would be five parts and the frequency of sampling every hour. So at 8:00 A.M., after one hour of production, the operator pulls five parts and measures the outside diameter of each product. The five readings are written at the bottom of the chart, below 8:00. The readings are 40, 50, 55, 45, and 50.

In the second hour the operator pulls five more products from production, measures the same quality characteristic, and records the data on the chart below 9:00. This continues until the end of the shift at 14:00 (2:00 P.M.). The next shift operator restarts the process and at 15:00, pulls five more parts, measures, and records the data on the chart. The graveyard shift operator restarts the process at 22:00 (10:00 P.M.) and repeats the measurements. The day shift operator ends production at 8:00 A.M. By this time, 25 samples have been pulled from production and the chart is complete.

Calculate Average $\overline{X}$ and Range of Each Subgroup Sample. Once the five measurements have been written down on the chart, the average $\overline{X}$ and the range R are calculated for each subgroup sample. The results are written on the chart at sum, average $\overline{X}$, and range R.

The equations for calculating the average and the range of each subgroup are

$$\overline{X} = \frac{X_1 + X_2 + \cdots + X_n}{n}$$

$$R = X_{\text{highest}} - X_{\text{lowest}}$$

where $X_1, X_2, \ldots, X_n$ = measurements of subgroup samples
n = number of measurements in subgroup sample

In our example the sum, average, and range of the first subgroup sample are, respectively, 240, 48, and 15. These are calculated as follows:

Sum	$40 + 50 + 55 + 45 + 50 = 240$
Range	$55 - 40 = 15$
Average	$\frac{240}{5} = 48$

These and subsequent data for the subgroup samples are entered on the graph.

Select Scales for $\overline{X}$ and R Charts. In the blank $\overline{X}$ and R charts the scales are not labeled. There are no strict guidelines with regard to labeling or how scales should be incremented, but the charts should be constructed so that most data points fall within the center of the graph.

The vertical scales on the $\overline{X}$ and R charts can be incremented based on the following rules of thumb. The upper boundary is one-half the difference between the highest and lowest individual measured values added to the highest value. The lower boundary is one-half the difference between the highest and lowest individual measured values subtracted from the lowest value. This rule can be used in both charts. If the lower boundary is a negative number, it becomes 0.

In our example, the highest individual measurement is 80 mm taken at 15:00, and the lowest average value is 35 mm taken at 19:00 and 20:00 hours. The difference between these two numbers is 45. Half of 45 is 22.5; and rounded to the nearest whole number it is 23. This is added to the highest individual measurement of 80 and is subtracted from the lowest individual measurement of 35. The resulting upper boundary on the chart is 103 ($80 + 23 = 103$) and the lowest is 12 ($35 - 23 = 12$). Entering the upper and lower boundaries on the chart ensures that all values can be plotted on the chart. Instead of using individual measurements, subgroup sample averages, $\overline{X}$, can also be used to increment the $\overline{X}$ graph.

On the R chart the lowest value is often 0 and the highest value is about 1.5 times the largest range value. So if the largest range value is 35, the upper boundary for the chart is 52 ($35 \times 1.5 = 52$) and the lowest value is 0.

As seen in Figure 4.5, points in the middle of the charts are compressed in order to fit in the out-of-control points. Once these points have been eliminated, the limits of both charts can be recalculated and the variation of the points will be more pronounced. Also, if the sample size consists of six or less parts, the lower control limit LCL_R of the range is 0.

Plot Values. Once products have been measured, subgroup sample averages and ranges are calculated and plotted on the $\overline{X}$ and R charts. The points are connected by a solid line.

Calculate Average Range and Process Average. The process average is denoted by $\overline{\overline{X}}$, which is the average of the subgroup sample averages. The average range $\overline{R}$ and the process average $\overline{\overline{X}}$ are calculated using the following formulas:

$$\overline{R} = \frac{R_1 + R_2 + \cdots + R_k}{k}$$

where $\overline{R}$ = average range
$\quad R_1$ = first subgroup range
$\quad k$ = number of subgroups

$$\overline{\overline{X}} = \frac{\overline{X}_1 + \overline{X}_2 + \cdots + \overline{X}_k}{k}$$

where $\overline{\overline{X}}$ = process average, or average of the averages
$\quad \overline{X}_1$ = first subgroup average
$\quad k$ = number of subgroups

In our example, the average range $\overline{R}$ and the process average $\overline{\overline{X}}$ are calculated as follows:

$$\overline{R} = \frac{15 + 10 + 15 + 10 + \cdots + 15}{25}$$

$$= 14.4$$

$$\overline{\overline{X}} = \frac{48 + 49 + 49 + 54 + \cdots + 49}{25}$$

$$= 49.52$$

Calculate Trial Control Limits. Next the trial control limits are calculated to determine the limits of any random, chance variation. The control limits are based on the subgroup sample size and the variation in the ranges. Control limits are $\pm 3\sigma$ limits of the normal curve representing the subgroup sample averages, not the distribution of individual values. The distribution of the individual measured values is wider and is explained in the section on capability, later in this chapter.

The following formulas are used for calculating the control limits:

$$\text{UCL}_R = D_4 \times \overline{R}$$
$$\text{LCL}_R = D_3 \times \overline{R}$$
$$\text{UCL}_X = \overline{\overline{X}} + A_2 \times \overline{R}$$
$$\text{LCL}_X = \overline{\overline{X}} - A_2 \times \overline{R}$$

where $\overline{R}$ = average of ranges
$\quad \overline{\overline{X}}$ = process average, or average of the averages
$\quad A_2$ = $\overline{X}$-chart upper and lower control limit constant

$D_4 = R$-chart upper control limit constant
$D_3 = R$-chart lower control limit constant

In these formulas, D_4, D_3, and A_2 are constants used in calculating control limits. These constants are determined by the subgroup sample size. For example, the constants used to calculate the range upper and lower control limits with a subgroup size of 5 are 2.11 and 0, respectively. The constant used to calculate the $\bar{X}$ upper and lower control limits for a subgroup of 5 is 0.58.

Constants for several sample sizes are shown below:

n	2	3	4	5	6	7	8	9	10
D_4	3.27	2.57	2.28	2.11	2.00	1.92	1.86	1.82	1.78
D_3	—	—	—	—	—	0.08	0.14	0.18	0.22
A_2	1.88	1.02	0.73	0.58	0.48	0.42	0.37	0.34	0.31

Note that for sample sizes below 7, $D_3 = 0$.

The trial control limits of our problem are calculated as follows. Given $\bar{R} = 14.4$, $\bar{\bar{X}} = 49.52$, and a subgroup size of 5,

$$UCL_X = \bar{\bar{X}} + A_2 \times \bar{R}$$
$$= 49.52 + 0.58 \times 14.4 = 57.87$$
$$LCL_X = \bar{\bar{X}} - A_2 \times \bar{R}$$
$$= 49.52 - 0.58 \times 14.4 = 41.17$$
$$UCL_R = D_4 \times \bar{R}$$
$$= 2.11 \times 14.4 = 30.4$$
$$LCL_R = D_3 \times \bar{R}$$
$$= 0 \times 14.4 = 0$$

Construct Chart. The average range $\bar{R}$ and the process average $\bar{\bar{X}}$ are drawn as solid horizontal lines on the respective graphs. The control limits for both graphs are drawn as dashed horizontal lines. If specification limits are drawn, they are drawn as solid horizontal lines. Often they are omitted from a working $\bar{X}$ chart since the critical issue is whether plotted points are random and inside the limits.

The upper control limit $UCL_X = 57.87$ and the lower control limit $LCL_X = 41.17$ are shown on the $\bar{X}$ chart; the upper control limit $UCL_R = 30.4$ and the lower control limit $LCL_R = 0$ are given on the R chart in Figure 4.5.

Interpret Trial Limits. If a value is outside the control limits, an assignable cause is forcing the variation. For example, in Figure 4.5 subgroup sample 8 is outside the upper control limit UCL_X on the $\bar{X}$ chart and subgroup sample 18 is outside the upper control limit UCL_R on the R chart.

Once trial limits have been calculated, any out-of-control conditions in either chart are analyzed to determine the reason for the assignable cause and to prevent any recurrence in both charts. Out-of-control conditions were discussed earlier in this chapter.

In our example, assignable causes were identified rather easily. At 15:00 (3:00 P.M.) a new operator coming on shift was instructed how to machine the part. During this period, measurements in subgroup sample 8 were out of control on the high side of the $\overline{X}$ chart. It took one hour for the new operator to learn how to maintain the machine on the target value. Then at 1:00 A.M. there was a power bump, which affected the target setting of the machine. One product had an O.D. measurement of 75 mm, which caused subgroup sample 18 to go out of control on the high side of the R chart.

Once out-of-control conditions have been eliminated, process average and control limits are recalculated for both charts based on the remaining number of subgroups, which in this example is 23. Given the new values of $\overline{R} = 13.7$, $\overline{\overline{X}} = 48.13$, and a subgroup size of 5,

$$UCL_X = \overline{\overline{X}} + A_2 \times \overline{R}$$
$$= 48.13 + 0.58 \times 13.7 = 56.08$$
$$LCL_X = \overline{\overline{X}} - A_2 \times \overline{R}$$
$$= 48.13 - 0.58 \times 13.7 = 40.18$$
$$UCL_R = D_4 \times \overline{R}$$
$$= 2.11 \times 13.7 = 28.9$$
$$LCL_R = D_3 \times \overline{R}$$
$$= 0 \times 13.7 = 0$$

If the recalculated control limits are closer together, points previously inside the old control limits may now be outside the new limits. These new out-of-control points have to be identified, eliminated, and the limits recalculated again. If there are no more assignable causes present, all out-of-control conditions have been eliminated and the process is considered in control.

BASIC QUESTIONS

The $\overline{X}$ and R charts provide real-time information about a process. The charts display process information as parts are produced. So if there are abnormal conditions or patterns, the operator can adjust the process as changes appear, not afterward when defective parts may have been produced. The charts are documents for monitoring and analysis. If they are not used, measurements will be taken, results will be posted, and the process will not improve.

The following questions should be addressed by anyone implementing process control: What do the centerlines of both charts say about the process? Is the process in control or out of control? Is the process capable?

What Do the Centerlines of Both Charts Say about the Process? Both charts have centerlines. They are the averages of the recorded mean and range measurements. The ideal is to have the centerline on the target, or nominal, dimension. This means that there is no drift in the process and it is centered in the middle of the specification limits.

The centerline of the $\overline{X}$ chart is midway between the upper control limit UCL_X and the lower control limit LCL_X. The centerline of the R chart is midway between the upper and lower control limits UCL_R and LCL_R, unless the lower control limit is 0. In this case, the centerline is closer to the lower control limit.

If the process is not well centered on the $\overline{X}$ chart, the question should be asked, can it be centered at the target dimension easily? Depending on the type of process, centering may only entail a slight adjustment. For example, if a machine is reducing the outside diameter of a solid bar, then the operator just tweaks an adjustment and the process is centered on the target dimension. In some chemical processes involving complex interaction of parts, however, centering can involve adjusting several variables.

Is the Process in Control? A process in statistical control has all its points inside the control limits, and there is no unnatural variation inside or outside the charts.

If the process is not in control, the operator or process engineer must isolate, identify, analyze, and eliminate the cause of variation. It is possible that the immediate cause can be eliminated easily, while eliminating the root cause may take longer.

Is the Process Capable? A process in control should have its control limits well inside specification limits. This means manufactured parts conform to specifications. It is important to remember that control limits are calculated from subgroup samples. These are tighter, closer together than the process spread. If a process is in control and its control limits are near or extend beyond the specification limits, defective parts are being produced. This is unacceptable and the process is not considered capable.

Several options are available. A new machine may be purchased to produce parts to specification. Or engineering may be asked to evaluate the unattainable specifications and perhaps make them broader.

CHART ANALYSIS (PROCESS IN CONTROL)

A control chart identifies process changes as they occur, and if there are unnatural deviations, corrective action can be taken to eliminate the cause of the changes. A process is in control when assignable causes have been removed and all nonrandom measured points eliminated. A process in control is assumed to have its measured average subgroup values distributed in a bell-shaped curve. A bell curve and a process in control have the following characteristics:

- ☐ No points are outside the control limits.
- ☐ Points are randomly distributed about the process average.
- ☐ About two-thirds of the points are located near the process average.
- ☐ Few points are near the control limits.

What If?

What if one outlying point has been identified and investigated but no assignable cause can be identified? It is possible that a point beyond the control limits is a false signal. This should be investigated. But if all causes have been considered, there may be no problem and the outlying point is natural and random.

Adjustment for no valid reason causes more injury than leaving a process alone. If control limits are already tight because of continuing process improvements, adjustment only increases rather than decreases variation.

Continuous Improvement

A process is in control when assignable causes have been eliminated so that all measured points are randomly distributed and are inside the control limits. The long-term goal is to center a process at the target specification and to eliminate assignable cause variation and to minimize any common-cause variation. For continuous improvement, a process may be improved by the purchase of a new machine, by additional personnel training, more uniform material from suppliers, or new methods of operation.

Figure 4.6 illustrates the use of an $\overline{X}$ chart to improve a process continuously. In this example the specification limits do not change over time. However, the measured data seem to center in the middle of the specification, as shown in the figure. Also, control limits move closer together over time.

If the specification stays the same, the process average coincides with the nominal value of the specification and the control limits have been tightened inside the specification limits; the chance of points appearing outside the specification limits is in terms of parts per million.

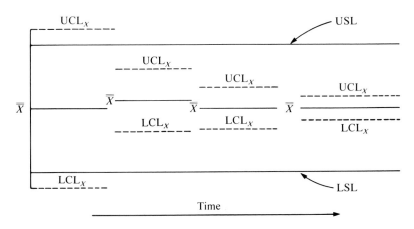

FIGURE 4.6
$\overline{X}$ chart illustrating continuous process improvement.

CHART ANALYSIS (PROCESS OUT OF CONTROL)

The $\overline{X}$ and R charts can be analyzed separately or together. Only one chart has to be out of control for the operator to search for an assignable cause. However, analyzing both charts provides more clues of assignable causes.

The assignable cause can be one or a combination of the factors discussed in the C–E diagram. The four patterns commonly found in one or both charts are sudden changes, runs, cycles, and multiple populations.

Sudden Changes. If there is a **sudden change,** or jump, in level, change is indicated in one or both charts (Figure 4.7). Possible assignable causes are:

☐ *Operator.* A new or inexperienced operator overadjusted a process, or the operator miscalculated control limits.
☐ *Material.* Raw material varies in physical or chemical properties.
☐ *Methods.* Procedures are different between shifts.
☐ *Machines.* A machine has a new fixture or tooling that changes the basic settings.
☐ *Environment.* The physical environment, such as an increase in humidity or contaminants, disrupts complex operations.

Runs. **Runs** are patterns, or trends, that occur inside or outside the control limits (Figure 4.8). Patterns outside the control limits obviously signal an out-of-control condition. Patterns inside the control limits are early warning signals that something unnatural is occurring and assignable causes are present. Change is seen in either or both charts.

A common indication that a process shift has occurred is having seven successive points above or below the process average, or seven successive points consistently rising or falling. Possible assignable causes are:

☐ *Operator.* A new inspector measures products differently.
☐ *Material.* Material is less uniform than usual.

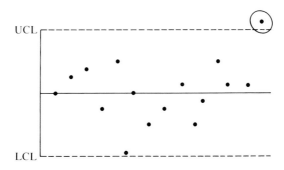

FIGURE 4.7
Control chart pattern indicating sudden changes.

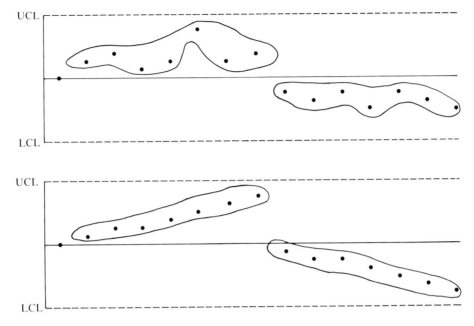

FIGURE 4.8
Runs indicating a process shift.

☐ *Methods*. Methods have changed over time to produce better or worse products. The trend line will increase or decrease.
☐ *Machines*. A fixture or die on the machine has loosened gradually, or the measurement equipment has changed.
☐ *Environment*. The dust or contaminants in a clean room have steadily increased and performance has deteriorated.

Cycles. Recurring **cycles** follow a pattern with periodic highs and lows (Figure 4.9). Change appears on both charts. Possible assignable causes are:

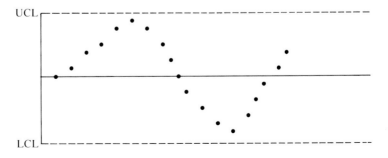

FIGURE 4.9
Change indicated by recurring cycles.

□ *Operator.* Different operators run a process differently.

□ *Material.* Raw material has changed or shows wide variations.

□ *Methods.* Operators in different shifts use different methods to process, inspect, or measure a product.

□ *Machines.* Process and inspection equipment in different shifts are different.

□ *Environment.* Environment changes because of temperature and humidity.

Multiple Populations. **Multiple populations,** that is, two or more populations, of data can show up on a chart (Figure 4.10). A continuous production run may take three shifts to complete. Each run is supervised by a different operator who uses different techniques.

If there are two populations of data, they show up as a bimodal histogram curve with widely distributed points in the $\overline{X}$ chart and wide variations in the R chart. Widely distributed points accumulating near UCL_X and LCL_X on an $\overline{X}$ chart can be thought of as the modes on a bimodal distribution. Possible assignable causes are:

□ *Operator.* Different operators use the same chart.

□ *Material.* Raw material has wide differences in composition.

□ *Methods.* Different operators use different methods to produce products.

□ *Machines.* One chart shows production from two machines.

□ *Environment.* Different shifts use the same chart, and humidity or temperature differences may affect measurements.

PROCESS CAPABILITY

Once it is known that a process is in statistical control, another important question is asked: Is the existing process or machinery capable of meeting customer specifications? This is called **process capability.**

Capability analysis is important to prevent defects from occurring. Once an engineer develops specifications, how does he or she know that the specification is realistic? In other words, how does the engineer know that the manufacturing process is capable of consistently making products whose dimensions are in the

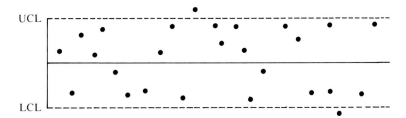

FIGURE 4.10
Control chart pattern indicating multiple populations.

middle of the specification spread? One way is to do a trial run using a process and to check the product measurements against the specification. If the process is capable of making products that conform to the specification, the specification can remain the same. If the process is not capable of consistently making products to specification, several options are available. A new piece of equipment may be purchased. Another machine may be assigned the task of producing the product. Or the specification may be relaxed, that is, the specification spread is broadened.

Other uses of capability studies are: determining the percentage of defective parts that can be expected from a normal running process, preparing studies of the cost of scrap that may be expected from the process, and performing manufacturability studies.

Process Is in Control

Process capability is evaluated after assignable causes have been identified, monitored, and eliminated. The process is now in statistical control. When the **process spread** is compared with the specification spread and it is determined that specifications can be achieved and met consistently, the process is capable.

Statistical control assumes that the process measurements follow a bell-shaped normal distribution. Normality can be checked in various ways. A histogram can be constructed to see whether measurements follow a bell shape. Another method for checking normality is to graph points on probability paper. A statistics book will provide analytical methods for determining normality.

Distribution of Averages and Individual Values

To calculate process capability, a change in thinking is required. Up to now we discussed control limits in terms of a distribution of subgroup sample averages $\overline{X}$, but now we must think in terms of the distribution of individual values X.

We are no longer interested in the distribution of the average values, which were used to calculate control limits. We want to know how the distribution of individual values, the process spread, will compare with the specification spread. The distribution of individual values represents the population of manufactured products. The specification spread is shown on the engineering print and is determined by an engineer. The distribution of individual values X will always be wider than the distribution of the averages $\overline{X}$ (see Figure 3.9).

Calculating the Population Standard Deviation

The spread, or variation, for the distribution of individual values is calculated using the population standard deviation σ. The population standard deviation can be determined in two ways: it can be calculated using the standard deviation formula discussed in Chapter 3 or it can be estimated. Using the formula discussed in Chapter 3, the population standard deviation σ is

$$\text{Population standard deviation} = \sigma = \sqrt{\frac{\Sigma(X - \mu)^2}{N}}$$

where σ = population standard deviation
Σ = summation sign
X = observation value, individual measurement
μ = population mean
N = total number of observations or measurements in population

The population consists of all the products produced by the process. If the number of products produced is large, this calculation becomes complex because each product from the process must be measured and a process, such as a stamping operation, may run 1000 or more pieces.

Since the above calculation of population standard destination can become complex, the quality analyst would prefer to calculate sample standard deviation using the familiar formula:

$$s = \sqrt{\frac{\Sigma(X - \bar{X})^2}{n - 1}}$$

If 25 or more samples are pulled from the population, the above formulas produce similar results.

Estimating the Population Standard Deviation

The preferred option is to estimate σ. The estimated standard deviation is denoted by $\hat{\sigma}$. The formula uses $\bar{R}$, which is the average of the subgroup sample ranges for a period when the process was known to be in control; d_2 is a constant developed for different subgroup sample sizes.

The estimated process standard deviation $\hat{\sigma}$ is equal to the average range $\bar{R}$ divided by the constant d_2. Mathematically,

$$\hat{\sigma} = \frac{\bar{R}}{d_2}$$

where $\hat{\sigma}$ = estimated process standard deviation
$\bar{R}$ = average range
d_2 = constant

The average range $\bar{R}$ is calculated by pulling 25 or more samples from the process; it is the average of the subgroup sample ranges for the period when the process was in control. The constant d_2 is selected from the following table. It varies with the sample size.

Number of Observations in Subgroup	2	3	4	5	6	7	8
d_2	1.128	1.693	2.059	2.326	2.534	2.704	2.847

C_p

Before describing how to calculate a **capability index,** several key terms must be reviewed. USL is the upper specification limit, LSL is the lower specification limit. The distance between USL and LSL is the specification spread, which is also called specification width or tolerance width. The midpoint of the specification is the target, or nominal, value.

The capability index, denoted by C_p, is the ratio of the specification spread over the process spread. The process is in statistical control, so the process spread is equal to six process standard deviations, $6\hat{\sigma}$. The process capability is calculated as follows:

$$C_p = \frac{\text{USL} - \text{LSL}}{6\hat{\sigma}}$$

where C_p = capability index
   USL = upper specification limit
   LSL = lower specification limit
   $\hat{\sigma}$ = estimated process standard deviation

In our previous example with the process in control there were five observations in each subgroup sample. The process capability is then calculated using the revised data as follows:

$$\hat{\sigma} = \frac{\overline{R}}{d_2}$$

$$= \frac{13.7}{2.326} = 5.89$$

Hence,
$$C_p = \frac{\text{USL} - \text{LSL}}{6\hat{\sigma}}$$

$$= \frac{70 - 30}{6 \times 5.89} = 1.13$$

Three situations of C_p can result: $C_p = 1$, $C_p > 1$, or $C_p < 1$ (Figure 4.11).

$C_p = 1$. This is illustrated by Figure 4.11(a). A bell curve, representing the normal distribution of the population of products, is placed between the upper and lower specification limits. This bell curve represents the process spread.

A capability of 1 ($C_p = 1$) means that the width of the specification spread is equal to the width of the process spread. The curve is furthermore assumed to be centered at the target, that is, in the middle of the specification spread. The distance between the upper and lower specification limits and the capability limits is $6\hat{\sigma}$. As you may recall, if capability equals 1, then the number of products between $\pm 3\sigma$ is 99.7% of the population. This means that the defective rate is 3 parts out of every 1000.

$C_p < 1$. If the capability index is less than 1, the process spread is greater than the specification spread. The bell curve is flatter, showing that there is more spread, or dispersion, of data. The curve is again centered in the middle of the specification spread. The defective rate is now greater than 3 out of 1000 parts [Figure 4.11(b)].

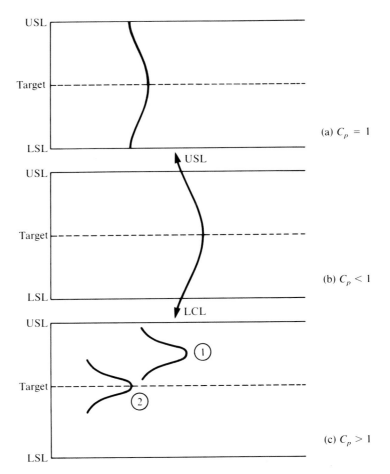

FIGURE 4.11
Capability indexes.

$C_p > 1$. If the capability index is greater than 1, the process spread is less than the specification spread. There is now less variation, and a narrower bell curve with a higher peak indicates this condition. If the capability spread is centered, as in Figure 4.11(c), and sufficiently narrow, the defective rate conceivably can be as low as parts per million.

C_{pk}

The C_{pk} index is another technique for calculating capability. It does not assume that the curve is centered in the middle of the specification spread. A curve may not be centered but may still be producing defective parts.

Capability can be described more accurately in terms of how well centered the curve is in the specification spread and how tight the variation. Figure 4.12

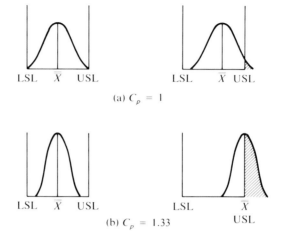

FIGURE 4.12
Curves, centered and off center.

shows curves for $C_p = 1$ and $C_p = 1.33$. According to the C_p index, these curves are capable as long as they are well centered within the specification spread. However, if a curve shifts to the right, defective units (shaded area) are produced. So the C_{pk} index was created to account for how well a curve is centered.

C_{pk} is calculated in a similar way as C_p. Variations within the subgroup sample are again indicated by $\overline{R}$ values. Since the process is in control, the process standard deviation can be estimated by calculating $\hat{\sigma}$ as before.

In the example worked out earlier in this chapter, $\overline{R} = 13.7$ and $d_2 = 2.326$. Then

$$\hat{\sigma} = \frac{\overline{R}}{d_2}$$

$$= \frac{13.7}{2.326} = 5.889$$

Also, $\overline{\overline{X}} = 48.13$, USL = 70, and LSL = 30. C_{pk} is the smaller of the following:

$$C_{pk} = \frac{\text{USL} - \overline{\overline{X}}}{3\hat{\sigma}} = \frac{70 - 48.13}{3 \times 5.889} = 1.237$$

and

$$C_{pk} = \frac{\overline{\overline{X}} - \text{LSL}}{3\hat{\sigma}} = \frac{48.13 - 30}{3 \times 5.889} = 1.027$$

Thus

$$C_{pk} \text{ (min)} = 1.027 \approx 1.03$$

The smaller of the two C_{pk} values is the true C_{pk}. Hence $C_{pk} = 1.03$ while $C_p = 1.13$. This means that the curve is centered at 48.14 and not at 50.00, which would be the middle of the specification spread. If the curve were centered in the middle of the specification spread, the theoretical C_{pk} value would be 1.13, which is closer to the C_p value.

Improving Capable Processes

If the process cannot achieve specifications, it can only be improved through management action. Management action means focusing on eliminating the chance or common causes of variation. W. Edwards Deming,* a quality guru, believes that management controls 85% of process improvement, while the operator controls 15%.

Once assignable causes have been eliminated, management looks at other factors that may cause variation, as discussed in connection with the C–E diagram. Machine performance may be improved through purchase of a new machine. Material uniformity may be improved by requiring vendors to implement statistical process controls. Operator variation may be improved through additional training. Specifications and engineering drawings may be changed to permit a machine or process to attain capability.

Let us examine the option of purchasing a new machine that can maintain capability. Once a machine is set at a target value, the only deviation should be from common causes. However, if the machine is old or has been abused, its repeatability, that is, its ability to maintain the target dimension, will be low. If a particular dimension is critical to safety and the specification limits cannot be relaxed, then the option is to purchase a new machine that is capable of maintaining the dimension with no assignable deviation. The decision of purchasing a new machine has to be analyzed in terms of costs and benefits.

INDIVIDUAL MEASUREMENT CHART

Most variable charts are based on three or more readings that form a subgroup sample. The operator averages and calculates the range in each subgroup sample. However, this may not be possible when measurements are expensive or difficult

---

* W. E. Deming, *Out of the Crisis* (M.I.T. Press, Cambridge, MA, 1986).

to perform. For instance, each measurement may require destruction of the part. So control charts for individual measurements were developed.

The **individual measurement chart** is quick and easy to construct. Its ease must be balanced against its lack of accuracy. It is not as sensitive to piece-to-piece variations as the $\overline{X}$ and R charts are. Since the chart plots one point, the range values can show substantial variation.

Procedure

The procedure for constructing an individual chart is similar to the $\overline{X}$ and R charts (Figure 4.13):

- ☐ Use a blank $\overline{X}$ and R chart.
- ☐ Record one reading on the chart.
- ☐ Calculate the moving range from individual measurements.
- ☐ Select the scales.
- ☐ Plot the measurement points.
- ☐ Calculate the control limits.
- ☐ Interpret the chart.

The individual value chart is similar to the $\overline{X}$ and R charts. However, since the range is a moving value, there is one less range value than there are X values. The constants D_4 and D_3 used in calculating the range control limits are the same as in the $\overline{X}$ and R charts, but the constant A_2 is replaced by E_2. The constants also vary according to the sample size used in grouping the moving ranges. In our example, the grouping is 2. The constants are given in the following table:

n	2	3	4	5	6	7	8	9
D_4	3.27	2.57	2.28	2.11	2.00	1.92	1.86	1.82
D_3	—	—	—	—	—	0.08	0.14	0.18
E_2	2.66	1.77	1.46	1.29	1.18	1.11	1.05	1.01

The control limits for an individual chart are calculated as follows:

$$\text{UCL}_R = D_4 \times \overline{R}$$
$$= 3.27 \times 1.26 = 4.12$$
$$\text{LCL}_R = D_3 \times \overline{R} = 0$$
$$\text{UCL}_X = \overline{\overline{X}} + E_2 \times \overline{R}$$
$$= 7.4 + 2.76 \times 1.26 = 10.75$$
$$\text{LCL}_X = \overline{\overline{X}} - E_2 \times \overline{R}$$
$$= 7.4 - 2.66 \times 1.26 = 4.05$$

Interpretation

The individual chart is interpreted in a similar manner as the $\overline{X}$ and R charts. Any point above or below the control limits or any nonrandom pattern indicates an

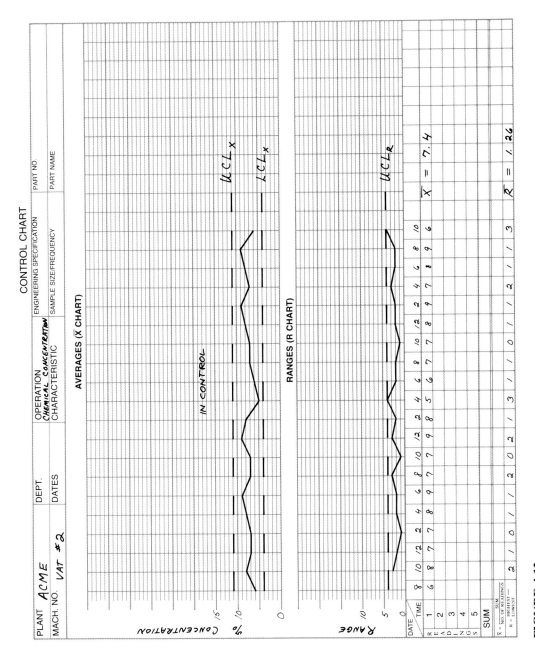

FIGURE 4.13
Individual measurement chart.

101

assignable cause. These causes are analyzed and eliminated in the same way as for the $\overline{X}$ and R charts. Once capability is thus established, management tries to eliminate chance causes

SUMMARY

☐ Statistical process control is a method for monitoring and controlling a process instead of inspecting products.

☐ SPC is based on the following assumptions:
Quality is conformance to specifications.
Processes and products vary.
Variation in processes and products can be measured.
Normal variation follows a bell-shaped curve.
Variation due to assignable causes distorts bell curve.
Variation is controlled through SPC.

☐ There are two types of data: attribute and variable. Attribute data are countable. Examples include conforming/nonconforming, go/no-go. Variable data are measurable by incremental amounts. For example, a ruler measures variable data.

☐ There are two types of causes: chance and assignable. Chance causes, also called common causes, are random causes, which are a natural part of the system. Chance causes usually cannot be detected or measured. Assignable causes, also called special causes, are not random. Assignable causes can be detected, measured, controlled, and eliminated.

☐ Variable chart, also called $\overline{X}$ and R chart, is composed of two charts, an $\overline{X}$ chart and an R chart. The $\overline{X}$ chart measures the average, or mean, of samples of measured parts. The R chart measures the range, or dispersion, of the data in the samples of measured parts.

☐ $\overline{X}$ and R charts are used to monitor machines, gather information about manufacturability, supply information to customers, and provide a visual baseline for process improvement.

☐ A process can be in control or out of control. A process in control has natural variation and plotted points are random. Natural variation is caused by chance factors. A process out of control has unnatural variation, which can be recognized and eliminated. Unnatural variation is caused by assignable factors.

☐ Specification limits and control limits are different concepts. Specifications are established by engineers for one product. Control limits are calculated from samples of measured data.

☐ SPC must be planned carefully, otherwise data will not be used. SPC planning starts with obtaining management support, defining the production system, checking the specifications, identifying product characteristics, determining locations that can cause variation, defining a measurement system, and focusing on continuous improvement.

☐ First a trial variable chart is constructed if it is not known whether the process is in control or not.

☐ The steps followed for constructing an $\overline{X}$ and R chart are: obtain data, choose subgroup sample size, calculate $\overline{X}$, select scales, plot values, calculate average range, calculate process average, calculate trial control limits, construct chart, and interpret trial limits.

☐ Once a chart is constructed, it is analyzed in terms of whether it is in control or not in control. A process is in control when assignable causes have been eliminated so that all measured points are inside the control limits and nonrandom patterns do not exist. A process in control only has chance causes. A process is out of control when assignable factors cause points to be outside the upper or lower control limits or when there are nonrandom patterns inside or outside the control limits.

☐ Out-of-control patterns can be seen in either the $\overline{X}$ or the R chart, or both. Four assignable causal patterns are sudden changes, runs, cycles, and multiple populations. The causes for these patterns are the main categories of the C–E diagram: operator, material, methods, machines, and environment.

☐ Process capability is a study to determine whether specification limits can be attained by a process or machine. Capability can be defined by C_p and C_{pk}. A capability index of 1, $C_p = 1$, means that the width of the specification spread is equal to the stabilized process spread. C_{pk} is another index of capability, which measures the spread and the centering of the distribution spread of points.

☐ Capable processes can be improved through management action. When a process is capable, assignable causes have been eliminated. Only management can improve a process through changing procedures, purchasing a new machine, or authorizing additional training of personnel.

☐ Individual measurement charts are variable charts that measure individual measurements, not the average of sample measurements. Individual measurement charts are constructed because measurements may require the destruction of parts, which is expensive.

KEY TERMS

assignable cause   Factor that contributes to variation; can be identified and eliminated.

attribute data   Qualitative data that can be counted; usually classified as good/bad, go/no-go.

average   Sum of values divided by the number of values; denoted by a bar over the symbol of values.

capability   State when the process average plus or minus the 3σ spread of the distribution of individual measurements is inside the specification tolerance spread.

capability index   Measure of capability of a process.

centerline   Line on a control chart that represents the long-term average of the measure being plotted.

chance cause   Random factor that contributes to variation; cannot be detected or identified; also called common cause.

common cause   Same as chance cause.

control chart   Graphic method for determining whether a process is in a state or statistical control.

control limits   Upper and lower limits on the bell curve, representing $\pm 3\sigma$ of the subgroup sample averages; used as criteria for signaling need for action by process operator; used as basis for judging the significance of variation from subgroup sample to subgroup sample.

cycles   Recurring nonrandom pattern with periodic highs and lows appearing on $\overline{X}$ and/or R charts.

individual measurement chart   Variable chart based on individual measurements.

lower control limit (LCL)   Control limit for points below centerline of curve.

micrometer   Measuring instrument that looks and works like slide rule; used to measure inside, outside, and depth dimensions.

multiple populations   Two or more populations of data appearing on $\overline{X}$ and/or R charts.

out of control   State where assignable factors are causing distortion in a chart.

process   Individual or a combination of factors, including operator, machine, material, methods, and environment, that produce a product or service.

process capability   Study to determine whether process is able to meet specifications consistently.

process spread   Extent to which the distribution of individual measurements of a quality characteristic varies.

quality planning   Sequential steps in implementing SPC.

R chart   Control chart for evaluating the dispersion or variation within a subgroup sample; also called range chart.

range   Difference between highest and lowest values in a subgroup sample.

run   Pattern or trend occuring inside or outside of control limits indicated on $\overline{X}$ and/or R charts.

sample size   Number of units in a sample.

specification limits   Same as tolerance limits.

subgroup sample   One or more measurements used to analyze a machine or process.

sudden change   Jump in level indicated on $\overline{X}$ and/or R charts.

tolerance limits   Acceptable dimensional boundaries for an individual product; dimensions inside tolerance limits are acceptable; also called specification limits.

trial control limits   Control limits calculated to determine any assignable variation.

upper control limit (UCL)   Control limit for points above centerline of curve.

variable control chart   Chart composed of two graphs, $\bar{X}$ and R, measuring central tendency and dispersion of data from a process.

variable data   Data that can vary by increments.

variation   Spread or dispersion.

$\bar{X}$ chart   Control chart for evaluating differences between averages of samples pulled from a process; also called average chart.

QUESTIONS

1. Discuss and explain the fundamental assumptions behind SPC.

2. What are the differences between attribute and variable data? Given an example of each.

3. Explain the statement: Attribute data do not provide specific information for improving a process.

4. What is the difference between change and assignable causes?

5. If a process has a tight distribution spread but has moved to the lower specification limit and is producing nonconforming parts, what should the operator do to produce conforming parts?

6. Can assignable causes be eliminated from the system? Discuss and provide an example.

7. How are assignable causes monitored, controlled, and eliminated?

8. A variable control chart can be used for what purposes?

9. Sampling from a process detects sample-to-sample variation. Discuss.

10. What does "statistical control" mean?

11. What does "out of control" mean? What should the operator do to bring the process in control?

12. What does the $\bar{X}$ chart graph? What does the R chart graph?

13. What is the relationship between specifications and control limits?

14. Why should SPC be planned? Discuss the steps for implementing SPC.

15. What is the purpose of a trial chart?

16. The scales on the $\bar{X}$ and R charts are not incremented. What are rules of thumb for incrementing the $\bar{X}$ and R charts?

17. What is the purpose for calculating trial control limits? Will they be different than the in-control limits? Discuss.

18. Upper and lower control limits relate to what limits on a bell curve?

19. Discuss the possible out-of-control patterns in a variable chart.

20. Multiple populations are indicated on a chart by what type of pattern?

21. What causes runs on a variable chart?

22. What is process capability? What do $C_p = 1$, $C_p < 1$, and $C_p > 1$ mean?

23. What is the purpose of C_{pk}? When will $C_{pk} = C_p = 1$?

24. Can capable processes be improved and if so, how?

25. What are the advantages of the individual measurement chart and when should it be used?

PROBLEMS

1. The following quality data, in millimeters, were obtained from a production run. The run had a high scrap rate. The subgroup sample size was 5. Calculate the control limits and graph the data on an $\overline{X}$ and an R chart.

Subgroup	$\overline{X}$	R	Subgroup	$\overline{X}$	R
1	185	23	11	180	9
2	180	8	12	176	8
3	176	22	13	178	7
4	170	12	14	181	4
5	176	7	15	180	6
6	180	8	16	177	6
7	179	15	17	178	10
8	180	6	18	182	9
9	178	7	19	178	7
10	178	12	20	176	10

2. The following are results of a tensile test of steel specimens. The results are in units of 1000 psi. Construct an $\overline{X}$ and an R chart. The subgroup number is 5.

Subgroup	$\overline{X}$	R	Subgroup	$\overline{X}$	R
1	48.2	2.3	14	49.5	1.8
2	49.2	3.4	15	49.4	2.1
3	50.3	2.1	16	49.3	2.3
4	51.9	2.4	17	53.2·	2.5
5	49.4	1.9	18	50.8	2.1
6	48.0	2.0	19	50.4	2.3
7	49.6	2.3	20	49.2	2.5
8	48.9	2.0	21	51.1	2.3
9	49.5	2.2	22	49.6	2.1
10	51.2	2.1	23	50.2	1.6
11	51.0	2.1	24	51.3	0.9
12	50.3	1.9	25	51.5	1.5
13	49.7	1.8			

3. Recalculate Problem 1 using a subgroup sample size of 4.

4. Recalculate Problem 2 using a subgroup sample size of 4.

5. Calculate C_p and C_{pk} for Problem 1.

6. Calculate C_p and C_{pk} for Problem 2.

7. In a manufacturing process, $\overline{\overline{X}} = 182$ mm and $\overline{R} = 8$ mm. The subgroup sample size is 4. Product specifications are 180 mm ± 20 mm.
   a. Find the control limits for the $\overline{X}$ and R charts.
   b. How many nonconforming products are being produced assuming the distribution of the product is normal?
   c. Is the process capable of meeting specifications?
   d. If the process is centered at the target, what percentage of the products do not meet specifications?

8. Recalculate Problem 7 using a subgroup sample size of 5.

9. Product specifications are 150.0 mm ± 20.0 mm. The subgroup sample size is 4. After 50 subgroups, $\Sigma X = 8000$ and $\Sigma R = 900$.
   a. Find the control limits for the $\overline{X}$ and R charts.
   b. How many nonconforming products are being produced assuming the distribution of the product is normal?
   c. If the process is centered at the target, what percentage of the products do not meet specifications?

10. Recalculate Problem 9 using a subgroup sample size of 5.

5

Practical Probability

"There is a 50% chance of rain today." "The U.S. has one in three chances of winning the Olympic gold medal in the women's 100-meter freestyle swimming event."

These statements are **probabilities,** or estimates, of what may occur in the future. Estimates, or forecasts, are made on every event in our lives. These estimates are based on the likelihood or probability of an occurrence.

In this chapter probability is developed in terms of applied mathematical models based on simple definitions, theory, and applications. Rigorous proofs are left to statistics and probability books.

The study of probability is useful for several reasons. Probability helps managers make decisions on quality when the outcome is uncertain. It is the theoretical basis for sampling, which is the main statistical technique in quality inspection.

INTRODUCTION TO PROBABILITY

Fundamental Definitions and Concepts

Probability and statistics use similar mathematical concepts but approach problems differently. In statistics, characteristics of a population of data are unknown. A sample is pulled from a population, and characteristics of the sample are known through measurement. Then inferences can be made of the whole population. In probability problem solving, the reverse occurs. Characteristics of the probability population are known, and the probability of obtaining a certain sample is unknown.

In probability theory, an event is an uncertain occurrence. For example, if a 5 is displayed when a die is tossed, the 5 is an **event.** If the die is thrown again, the 3 is another event. The activity producing the event is called an **experiment.** The toss of a coin or die is an example of an experiment.

The set of all possible outcomes of the experiment is called the **sample space** of the experiment. In the die toss experiment, the sample space has six members, which are the six faces of the cube.

Probability theory also uses mathematical symbols to present and explain ideas. The probability of event E occurring is expressed as $P(E)$. Probabilities are measured between 0.0 and 1.0. A 100% chance of an event occurring is 1.0. A 0% chance of an event occurring is 0.0. A 60% probability of an event occurring is 0.6.

The probabilities described in this chapter characterize independent events. If an event has no effect on future events, it is a **statistically independent event.** The concept of statistical independence is important to understanding sampling theory, which is introduced in Chapter 7.

What Is Probability?

There are three basic definitions, each leading to a different approach to probability: subjective, classical, and relative frequency.

Subjective Approach. In conversational usage, probability means the likelihood, or chance, that something will occur. This is called **subjective probability** because a personal evaluation is used to determine the long-run frequency, or occurrence, of an event. ''There is a 50% chance of rain today'' is an example of subjective probability. The person saying this is offering a personal appraisal. There is no evidence that the person has substantive knowledge. Thus the subjective approach is not precise.

Classical Approach. The **classical** approach to **probability** assumes that outcomes, or events, of an experiment have an equal chance of occurring. In this approach the probability of an event occurring is calculated by dividing the number of desired outcomes by the total number of possible outcomes. This approach is more rigorous and accurate than the subjective approach.

For example, a balanced die has six faces. If the die is tossed a large number of times, there is an even chance that any one of the six sides will come up. In the language of statistics, if the die is tossed a large number of times, a population of numbers 1, 2, 3, 4, 5, and 6 is generated. If the population is large, the probability of any value or event occurring is the same. For example, the probability of a 6 appearing on one toss of a die becomes one chance out of six or, expressed statistically,

$$P(6) = \frac{\text{number of desired outcomes}}{\text{total of possible outcomes}}$$

$$= \frac{1}{1 + 1 + 1 + 1 + 1 + 1} = \frac{1}{6}$$

Relative Frequency. The **relative frequency** approach is based on historical data. In other words, what happened in the past can be used to infer future trends, assuming historical patterns were stable. Thereby the relative frequency definition states that the probability of a future event occurring is the number of times the event occurred in the past divided by the total number of specific observations.

The relative frequency approach is used when there are a large number of possible events and the probability of any event occurring, such as a defective product, is small. The relative frequency approach can be used in manufacturing and service industries. In manufacturing, machines mass-produce products. In service industries, people and machines process large amounts of paper. In both examples, relatively few nonconforming products or documents are generated, and the probability of selecting a nonconforming product is not equally likely.

In the following section we cover the fundamental rules of unconditional, joint, and conditional probabilities. The rules look complicated because of the new terms. However, once these are understood, the rules of probability are relatively simple. We illustrate each of the rules by using simple set theory, probability trees, or Venn diagrams.

RULES OF PROBABILITY

Unconditional Probability

Unconditional probability is the probability of a single event occurring. For example, in any toss of a die, only one of the six faces of the die can come up, a 1, 2, 3, 4, 5, or 6. Unconditional probability is computed by:

$$P(E) = \frac{n}{N}$$

where $P(E)$ = probability of event E occurring
N = number of trials
n = number of observations

Mutually Exclusive Event. Unconditional probability is an example of a **mutually exclusive** event. Events are mutually exclusive if one and only one event can occur at the same time. Using the example of a die, only one face will come up in any toss.

The probabilities of two or more mutually exclusive events occurring can be expressed by the special rule of addition for mutually exclusive events. This rule states that if two events are mutually exclusive, the probability of either X or Y occurring is the sum of their individual probabilities:

$$P(X \text{ or } Y) = P(X) + P(Y)$$

The equation for three mutually exclusive events is

$$P(X \text{ or } Y \text{ or } Z) = P(X) + P(Y) + P(Z)$$

The addition rule for three mutually exclusive events X, Y, and Z is illustrated in Figure 5.1.

EXAMPLE   Products from machines are 100% inspected for quality. Out of 2000 parts, the machines have the following production history:

	Number of Defectives	Probability of Occurrence
Machine W	10	0.005
Machine X	5	0.0025
Machine Y	20	0.01
Machine Z	10	0.005

What is the probability that machines W and Z will produce defective parts?

Solution

$$P(W \text{ or } Z) = P(W) + P(Z)$$
$$= 0.005 + 0.005 = 0.01$$

Venn diagrams. Venn diagrams are useful for presenting and understanding simple probability concepts. They are either interlocking or separate circles.

Venn diagrams are easily constructed. First a rectangular space is drawn, representing all possible outcomes. This is called the sample space. Then the areas of the two circles, designated as event X and event Y, are added together. In set theory, a mutually exclusive relationship is called a union and can be illustrated by a Venn diagram, as shown in Figure 5.2(a).

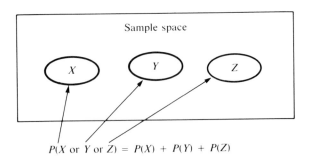

FIGURE 5.1
Addition rule for mutually exclusive events.

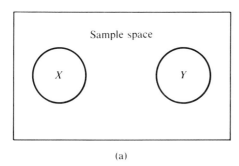 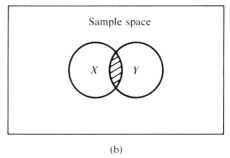

(a)   (b)

FIGURE 5.2
Venn diagrams. (a) Mutually exclusive events. (b) Not mutually exclusive events.

Not Mutually Exclusive Events. If two events are not mutually exclusive, both can occur at the same time. For example, a 10 and a diamond card pulled from a deck of cards are not mutually exclusive events. Both events can occur at the same time. It is possible that the 10 card is a diamond. Mathematically, if two events, or occurrences, are not mutually exclusive, the probability that either event or both may occur is expressed as follows:

$$P(X \text{ or } Y) = P(X) + P(Y) - P(X \text{ and } Y)$$

Again, a Venn diagram can express this formula [Figure 5.2(b)]. The event $P(X \text{ or } Y)$ consists of both circles. The event $P(X \text{ and } Y)$ is the crosshatched area where the two circles intersect. The event $P(X \text{ or } Y)$ is the area of the two circles minus the crosshatched intersection.

However, the probability of drawing a 10 of diamonds out of one pack of 52 cards is a mutually exclusive event. The 10 of diamonds is one card, and there is no possibility another 10 of diamonds can be pulled from the same deck. So the probability of this event is

$$P(10 \text{ of diamonds}) = \frac{1}{52}$$

The probability of drawing a 10 or a diamond card is not a mutually exclusive event. The following formula gives us the probability of this event:

$$P(10 \text{ or diamond}) = P(10) + P(\text{diamond}) - P(10 \text{ and diamond})$$
$$= \frac{4}{52} + \frac{13}{52} - \frac{1}{52} = \frac{16}{52} = \frac{4}{13}$$

Elementary set theory. The preceding probability rules can be illustrated by simple set theory. In the die toss experiment, the sample space is composed of six faces of the die and is expressed in set notation as

$$S = (1, 2, 3, 4, 5, 6)$$

If we define two subsets X and Y from this sample space, mutually exclusive examples can be illustrated:

$$S = (1, 2, 3, 4, 5, 6)$$
$$X = (1, 2, 3)$$
$$Y = (3, 5, 6)$$

Therefore

$$P(X \text{ and } Y) = 3$$
$$P(X \text{ or } Y) = 1, 2, 3, 5, 6$$

Joint Probability

If events X or Y are independent of each other, the probability of each occurring in succession is the product of their individual probabilities. This is called **joint probability** and is expressed symbolically as follows:

$$P(XY) = P(X) \times P(Y)$$

where $P(XY)$ = probability of events occurring in succession

$P(X)$ = probability of event X

$P(Y)$ = probability of event Y

Tossing dice in succession is an example of statistically independent events. The outcome of each toss is in no way associated with previous or subsequent events. When a six-sided die is tossed, the probability of any number coming up is

$$P(1, 2, 3, 4, 5, \text{ or } 6) = \frac{1}{6}$$

Furthermore, the probability of a 2 coming up in two successive tosses of one die is

$$P(2, 2) = \frac{1}{6} \times \frac{1}{6} = \frac{1}{36}$$

Probability tree. The probability of independent successive events can be illustrated by a **probability tree.** The probability trees for one toss and two tosses of a die are illustrated in Figure 5.3. The probability of each outcome is given by the probabilities of each path. In each toss there are six possible outcomes. Each has a probability of $\frac{1}{6}$. In two tosses, the probability of having two successive heads is $\frac{1}{36}$.

PERMUTATIONS AND COMBINATIONS

Permutations

This section discusses the rules of calculating permutations and combinations. A **permutation** is an ordered group of letters, objects, or products. For example, the possible three-letter permutations of the four-letter word TEAR are TEA, TAE,

Toss 1   Toss 2

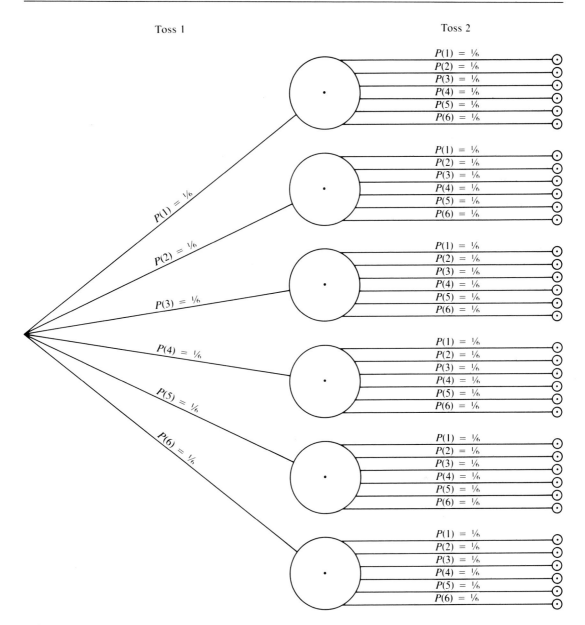

FIGURE 5.3
Probability tree based on two tosses of a die.

TER, TAR, TRA, TRE, EAT, ETR, EAR, ERA, ERT, ETA, ATE, ART, ARE, AET, ATR, AER, RAE, RTE, RTA, RAT, RET, and REA. Each group of three letters can be arranged into 24 permutations.

In order to calculate permutations and combinations, a new symbol is introduced, called the **factorial.** The mathematical symbol ! is the factorial sign. $N!$

means $n(n - 1)(n - 2) \cdots (0!)$. For example, 5! is 5(4)(3)(2)(1). By definition, 0! is equal to 1. The factorial is used to calculate large numbers.

The formula for calculating permutations in terms of factorials is

$$P_r^n = \frac{n!}{(n - r)!}$$

where P = number of ways objects can be arranged
n = total number of objects
r = number of objects used at one time

The calculation of the number of three-letter words made from the word TEAR can be simplified by using this formula. For $n = 4$ (total number of letters) and $r = 3$ (number of letters used),

$$P_3^4 = \frac{4!}{(4 - 3)!} = \frac{4!}{1!}$$
$$= 4 \times 3 \times 2 \times 1 = 24$$

This calculation becomes more complicated as the number of letters or objects becomes larger. For example, the number of five-letter permutations that can be made from the word ESTABLISHMENT is calculated as follows. For $n = 13$ (total number of letters) and $r = 5$ (number of letters used),

$$P_5^{13} = \frac{13!}{(13 - 5)!} = \frac{13!}{8!}$$
$$= \frac{13 \times 12 \times 11 \times 10 \times 9 \times \cancel{8 \times 7 \times 6 \times 5 \times 4 \times 3 \times 2 \times 1}}{\cancel{8 \times 7 \times 6 \times 5 \times 4 \times 3 \times 2 \times 1}}$$
$$= 154,440$$

Combinations

A **combination** is an unordered group of letters, objects, or products. Unlike a permutation, the ordering of letters, objects, or products in a combination is not important. This means that the number of letter combinations made from a word is smaller than the number of permutations. For example, there are 24 three-letter permutations of the word TEAR. Each of these is ordered in a different way. However, there are only four three-letter combinations of letters, TEA, EAR, ART, and RTE; and there is only one four-letter combination to the word, TEAR.

The number of combinations of letters can be calculated by the following formula:

$$C_r^n = \frac{n!}{r!(n - r)!}$$

where C = number of all possible combinations
n = total number of objects
r = number of objects used at one time

To determine the number of letter combinations in the word ESTABLISH-MENT, the calculation becomes more complicated since the number of letters or objects becomes larger. For example, the number of five-letter combinations made from the word ESTABLISHMENT is

$$C_5^{13} = \frac{13!}{5!(13-5)!} = \frac{13!}{5!8!}$$
$$= \frac{13 \times 12 \times 11 \times 10 \times 9 \times \cancel{8 \times 7 \times 6 \times 5 \times 4 \times 3 \times 2 \times 1}}{5 \times 4 \times 3 \times 2 \times 1 \times \cancel{8 \times 7 \times 6 \times 5 \times 4 \times 3 \times 2 \times 1}}$$
$$= \frac{154{,}440}{120} = 1287$$

BINOMIAL PROBABILITY DISTRIBUTION

What Is a Probability Distribution?

An understanding of frequency distributions leads to probability distributions. **Probability distributions** are theoretical frequency distributions, which show how events are expected to vary given certain conditions. In the rest of this chapter we introduce three important probability distributions: binomial, hypergeometric, and Poisson.

The **binomial** probability **distribution** can be used to solve quality problems where a continuous flow of products is produced by a process or supplied by a vendor. These products can be evaluated in terms of attributes such as good/bad, acceptable/defective, or heads/tails. The importance and use of this distribution will become clear when attribute control charts are discussed in the next chapter.

The binomial probability equation describes possible outcomes when only two mutually exclusive events can occur, such as the toss of a coin. It should be used carefully. The following should be observed when using the binomial distribution:

☐ Two events, called a success or a failure, must be mutually exclusive.
☐ Collected data are counted in whole numbers, 0, 1, 2, ...
☐ The probability of a success or failure stays the same for each trial.
☐ Trials are independent of each other; the outcome of one trial does not affect that of the other.

Example of a Binomial Equation

Coins again are used to illustrate this distribution. In each coin toss there are exactly two possible outcomes, heads or tails. This is expressed as the sum of two terms, the binomial (heads + tails) or, abbreviated, $(h + t)$. Using this abbreviation, the generalized binomial probability distribution can be developed.

The expression $(h + t)^2$ expanded is $(h + t)(h + t)$, which in a straightforward algebraic multiplication results in $h^2 + 2ht + t^2$. If the binomial expression is raised to the nth power, $(h + t)^n$, it means the expression $(h + t)$ is multiplied by itself n times. The usefulness of the binomial equation may be illustrated by an example. A binomial raised to the fourth power expands to

$$(h + t)^4 = h^4 + 4h^3t + 6h^2t^2 + 4ht^3 + t^4$$

Each element and coefficient in this expression reveals information. Elements are h^4, h^3t, h^2t^2, ht^3, and t^4. Coefficients are the numbers in front of the expressions, specifically 4, 6, and 4. The equation indicates the number and type of possible combinations of heads and tails that result if a coin is tossed four times.

The expression h^4 represents the outcome of four heads. The expression h^3t corresponds to the outcome of three heads and one tail. The expression h^2t^2 represents the outcome of two heads and two tails. The expression ht^3 represents the outcome of one head and three tails. The expression t^4 represents the outcome of four tails.

The coefficient of each element gives the number of combinations of heads and tails that can result from four tosses. There are one way to obtain four heads, four ways to obtain one head and three tails, four ways to obtain one head and three tails, six ways to obtain two heads and two tails, and four ways to obtain four tails.

Binomial Distribution in Quality

The generalized binomial probability distribution can be used to calculate the probability of one of the above combinations occurring. The generalized equation of the binomial probability distribution is

$$P(r) = \frac{n!}{r!(n - r)!} (p)^r(q)^{n-r}$$

where n = number of trials
r = number of successes
p = probability of success on each trial
q = probability of failure = $1 - p$

The probability an event will occur is indicated by $P(r)$. There are only two possible outcomes when using the binomial equation, success/failure, heads/tails, conforming/nonconforming, or go/no-go. The number of occurrences is denoted by n. The binomial distribution can be used when the probability of a defective product is constant from sample to sample and the lot size is infinite.

The binomial formula is used in three ways in quality control: first to compute the probability that an exact number r of results will be obtained in n repetitions of an experiment; second to compute that r or more results will be obtained in n repetitions of an experiment; and third to compute that r or less results will be obtained in n repetitions. As can be seen in the example below, the number of successes, r, becomes the number of defective products. Probability of success, p, is the defect rate. The probability of failure is $1 - p$.

EXAMPLE   A vendor is supplying a product to a customer in shipments of 50. The supplier guarantees that each shipment will have no defective products. The history of inspection has shown a defect rate of 0.5% for each shipment.

1. What is the probability that a shipment of 50 products will have no defective products?
2. What is the probability that it will have one or less defective products?
3. What is the probability that if a lot has one or less products, it will be rejected?

Solution
The equation to be used is:

$$P(r) = \frac{n!}{r!(n-r)!} p^r q^{n-r}$$

In this example, $n = 50$, $r = 0$, $p = 0.005$, and $q = 1 - 0.005 = 0.995$.

1. The probability $[P(r)]$ of no defective products ($r = 0$) out of a shipment of 50 ($n = 50$) with a known defective rate ($p = 0.005$) is calculated as follows:

$$P(0) = \frac{50!}{0!50!} (0.005)^0 (0.995)^{50}$$

$$= (0.995)^{50} = 0.778$$

2. The probability it will have one defective product is:

$$P(1) = \frac{50!}{1!49!} (0.005)^1 (0.995)^{49}$$

$$= 0.196$$

Therefore, the probability it will have 0 or 1 defective products is:

$$P(1 \text{ or less}) = P(0) + P(1) = 0.974$$

3. The probability the above is rejected is:

$$P(\text{reject}) = 1 - P(1 \text{ or less})$$

$$= 1 - 0.974$$

$$= 0.026$$

EXAMPLE   Past experience indicates that 0.1% of the parts produced by a computer-controlled high-speed machine are defective. What is the probability that out of 20 parts selected at random, no parts will be defective?

Solution

$$P(r) = \frac{n!}{r!(n-r)!} p^r q^{n-r}$$

given $n = 20$, $p = 0.001$, $q = 1 - 0.001 = 0.999$, and $r = 0$,

$$P(0) = \frac{20!}{0!(20 - 0)!}\ (0.001)^0(0.999)^{20-0} = 0.980$$

HYPERGEOMETRIC PROBABILITY DISTRIBUTION

The **hypergeometric distribution** allows the operator or inspector to find the probability of acceptable products in samples taken without replacement from small populations, such as in batches or shipments of products. The binomial distribution is used when samples are taken from a population of items and are placed back into the population. However, in most sampling this is not normal. When products are sampled from a shipment of products, products are set aside and not sampled again. Also, if sampled products are tested destructively, it is impossible to place these products back into the shipment.

The basic principles of the hypergeometric distribution can be illustrated by an experiment where marbles are randomly selected from a box.* At first we demonstrate the usefulness of the hypergeometric distribution through the use of the probability tree. Then we calculate probabilities by using the hypergeometric equation.

A box contains seven black and three white marbles. Three marbles are randomly selected, one at a time, and are not replaced. We want to find the probability of obtaining three white marbles. Figure 5.4 shows the probability tree for this problem. The number of marbles remaining in the box after a selection is shown inside the square. The paths leading to the end of the tree, such as on the upper branch (W_1, W_2, W_3) are drawn as solid lines. At first there are seven black and three white marbles. In the first cut, the probability of obtaining a white marble is $\frac{3}{10}$. This is shown on the uppermost path line.

If the first marble is white, seven black and two white marbles remain. The conditional probability that the second marble is white, assuming the first marble is white, is $\frac{2}{9}$. Following this path to the top of the probability tree, it is seen that the joint probability of pulling three white marbles in three successive samples is $\frac{6}{720}$. This is calculated by applying the joint probability multiplication rule discussed earlier:

$$P(W_1, W_2, W_3) = \frac{3}{10} \times \frac{2}{9} \times \frac{1}{8} = \frac{6}{720} = \frac{1}{120}$$

Hypergeometric Equation

The probability tree diagram is obviously too complicated to use when the number of items becomes large. The hypergeometric distribution can then provide the desired probabilities.

---

* From L. Lapin, *Statistics for Modern Business Decisions* (Harcourt Brace Jovanovich, New York, 1982), pp. 570–573.

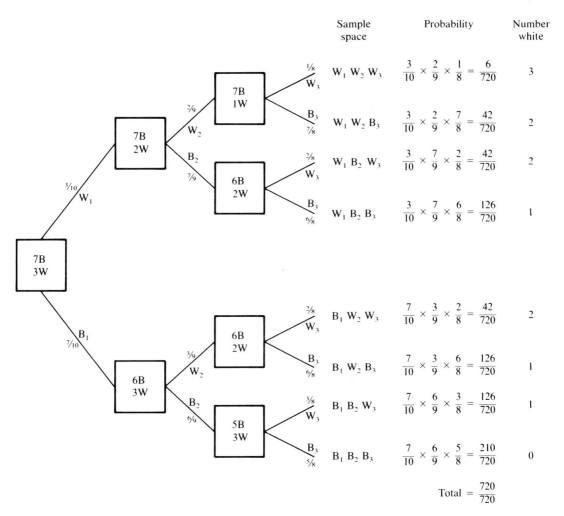

FIGURE 5.4
Hypergeometric probability tree.

Referring to the earlier example, the number of possible outcomes is the total number of combinations that can be formed of 10 marbles selected three at a time. This is expressed as C_3^{10}, which is the number in the denominator.

The number of ways to obtain three white marbles out of three is C_3^3. The number of ways to select zero white marbles out of 7 black ones is C_0^7.

Finally, the joint probability of obtaining three white marbles out of the total number of possible combinations is

$$P(\text{white marbles}) = \frac{C_3^3 C_0^7}{C_3^{10}}$$

In terms of factorials, this equation becomes

$$P(\text{white marbles}) = \frac{\dfrac{3!}{3!0!} \times \dfrac{7!}{0!7!}}{\dfrac{10!}{3!7!}} = \frac{3!7!}{10!}$$

$$= \frac{3 \times 2 \times 1 \times \cancel{7 \times 6 \times 5} \times \cancel{4 \times 3 \times 2 \times 1}}{10 \times 9 \times 8 \times \cancel{7 \times 6 \times 5} \times \cancel{4 \times 3 \times 2 \times 1}}$$

$$= \frac{6}{10 \times 9 \times 8} = \frac{6}{720} = \frac{1}{120}$$

EXAMPLE   ACME glass works produces four grades of crystal glass. Each grade is determined by the amount of hand-finished labor. The higher grades require more handwork and therefore sell for a higher price.

Each grade comes from a different production area. Shipping made a mistake and did not label the boxes according to grade. The products were placed in similar brown boxes as follows:

Grade	Number of Boxes
Best	1
Better	2
Good	7
OK	40

The best grade is sent to the highest paying customers. The better grade is sent to those who don't want to pay the highest price. Differences between best and better grades can only be seen through a magnifying glass. The highest paying customers will accept the two higher grades but reject the two lower grades. Shipping sent a customer three brown packages. The customer wanted three best or better grade products.

ACME wants to know the joint probability of the customer obtaining 1 best, 2 better, 0 good, and 0 OK grade packages.

Solution

$$P(1, 2, 0, 0) = \frac{C_1^1 C_2^2 C_0^7 C_0^{40}}{C_3^{50}}$$

$$= 1 \text{ chance in } 19,600$$

Hypergeometric Equation in Quality Control

The hypergeometric equation is used in quality control to calculate the probability of defective units in a sample from a population of products. The generalized hypergeometric equation calculates the probability of r successes in a random sample of n elements taken from a population N without replacement and is expressed as follows:

$$P(r) = \frac{C_r^S C_{n-r}^{N-S}}{C_n^N}$$

where $P(r)$ = probability of r successes
$\qquad\quad S$ = number of successes in population
$\qquad\quad N$ = total number in population
$\qquad\quad n$ = total number in sample
$\qquad\quad r$ = number of successes in sample
$\quad N - S$ = number of failures in population
$\quad n - r$ = number of failures in sample

The above equation can be thought in terms of:

$$P(r) = \frac{\left(\begin{array}{c}\text{combinations}\\\text{of failures}\end{array}\right)\left(\begin{array}{c}\text{combinations}\\\text{of successes}\end{array}\right)}{\text{combinations of all outcomes}}$$

Values in the generalized equation can ease the solution of a common quality control problem. The problem is to determine the probability of finding a certain number d of defectives in a sample of product taken from a shipment without replacement. The formula is then

$$P(d = r) = \frac{C_d^D C_{n-d}^{N-D}}{C_n^N}$$

where $P(d = r)$ = probability of d defectives
$\qquad\quad d = r$ = number of defectives in sample
$\qquad\quad D = S$ = number of defectives in population (lot)
$\qquad\qquad N$ = total number of products in population (lot size)
$\qquad\qquad n$ = total number of products in sample
$\quad n - d = n - r$ = number of conforming products in sample
$N - D = N - S$ = number of conforming products in population (lot)

Again, the above equation can be expressed as:

$$P(d) = \frac{\left(\begin{array}{c}\text{combinations of}\\\text{defective products}\end{array}\right)\left(\begin{array}{c}\text{combinations of}\\\text{conforming products}\end{array}\right)}{\text{combinations of all products}}$$

EXAMPLE   A shipment has 100 products. The supplier has a fraction defective of 0.02. What is the probability of drawing one defective product in a random sample of two?

Solution

$$P(d) = \frac{C_d^D C_{n-d}^{N-D}}{C_n^N}$$

Here $D = 2$, $N = 100$, $n = 2$, and $d = 1$. Then

$$P(d) = \frac{C_1^2 C_1^{98}}{C_2^{100}}$$

$$= \frac{\dfrac{2!}{1!1!} \times \dfrac{98!}{97!1!}}{\dfrac{100!}{98!2!}}$$

$$= \frac{2! \times 98! \times 98! \times 2!}{1! \times 1! \times 97! \times 1! \times 100!} = 0.0396 = 3.96\%$$

POISSON DISTRIBUTION

Many practical service quality problems involve events occurring over time. An important problem is how to design a facility in order to service customers arriving at unpredictable rates. If a manager knows at what rate customers arrive, management can determine the optimum number of people required to service the customers. For example, at peak demand times it may be necessary to bring on additional staff. This is called a queuing problem. Queuing problems are solved using a **Poisson distribution.** In analyzing arrival or departure patterns, the event is not as important as the number of occurrences. Other queuing examples include:

☐ Number of defectives in outgoing shipments
☐ Number of flaws in newly painted parts
☐ Arrivals of cars at a toll booth
☐ Optimum inventory levels
☐ Customers arriving at the bank
☐ Number of patients arriving at the hospital
☐ Snowstorms occurring in an area

The Poisson probability distribution is expressed mathematically as follows:

$$P(x) = \frac{e^{-\lambda}\lambda^x}{x!}$$

where $x = 0, 1, 2, \ldots$
e = base of natural logarithm = 2.7183
λ = mean rate at which events occur during process

The preceding examples of Poisson distributions share some common features. Using an example of customers arriving at a counter in a fast-food restaurant, the following occurs:

☐ The number of customers arriving at the counter is an integer: 0, 1, 2, 3, ...

☐ The average number of customers arriving at the counter can be estimated from past usage.

☐ The customers are approaching the counter randomly over time.

☐ The number of customers approaching the counter in any time interval is independent of what occurred in previous time periods.

☐ The process rate λ is constant during the period under investigation.

If the interval for an event to occur is relatively short, the probability of two or more events occurring in the same interval is rare. The Poisson distribution can be represented by a binomial distribution if only two outcomes are possible in each trial. (See Appendix B for Poisson distribution tables.)

Limitations

The Poisson distribution has several limitations. First, it assumes that the mean value of the occurrence rate is constant over an extended period of time. This may not be true. Often arrival or departure rates change over time because of random external interferences. Events can change because of shift, time of day, or other factors. For example, the people at a plant are paid on Fridays. The number of people who deposit their wages increases on Fridays after 5:00 P.M. So banks add more tellers to accommodate the increased demand and to avoid long lines.

EXAMPLE   Beth is the risk manager in a manufacturing company. She believes lost-time accidents follow a Poisson distribution. According to her analysis, there are about two lost-time accidents per month. If there is a high probability of a high number of accidents, the company must investigate its workman's compensation program and review its safety program. She wants to know the probabilities of having none, one, or six lost-time accidents per month.

Solution

$$P(x) = \frac{\lambda^x e^{-\lambda}}{x!}$$

Here $\lambda = 2$ accidents per month and $x = $ number of accidents. Then

$$P(0) = \frac{2^0 e^{-2}}{0!} = 0.1353$$

$$P(1) = \frac{2^1 e^{-2}}{1!} = 0.2707$$

$$P(2) = \frac{2^2 e^{-2}}{2!} = 0.2707$$

$$P(3) = \frac{2^3 e^{-2}}{3!} = 0.1804$$

$$P(4) = \frac{2^4 e^{-2}}{4!} = 0.0902$$

$$P(5) = \frac{2^5 e^{-2}}{5!} = 0.0361$$

$$P(6) = \frac{2^6 e^{-2}}{6!} = 0.0120$$

$$P(\text{total}) = P(0) + P(1) + P(2) + P(3) + P(4) + P(5) + P(6)$$
$$= 0.9954$$

Probability of no accidents $= P(0) = 0.1353 = 13.53\%$
Probability of one accident $= P(0) + P(1) = 0.1353 + 0.2707$
$$= 0.4060 = 40.6\%$$
Probability of six accidents $= 0.9954 = 99.54\%$

APPLICATION OF PROBABILITY DISTRIBUTIONS

The three probability distributions discussed in this chapter should be used under specific circumstances. The hypergeometric equation applies when a random sample is pulled without replacement from a finite lot with a fixed percentage of defective products. If the sample is pulled from a continuous stream of products with a fixed percentage of defective products, the binomial equation should be used. This equation will give the probability of whether the lot has r, r or more, or r or less defective products.

The binomial equation can be used in place of the hypergeometric equation if the ratio n/N is less than 0.1. However, using the hypergeometric or the binomial equation can be time consuming, as the example problems illustrate. A handheld statistical calculator has a factorial limit of 66! The binomial probability distribution for probabilities of success less then 0.05 ($p < 0.05$) can be calculated, but this is time consuming if the number of trials or the sample size is large ($n > 100$). Thus if p is small and n is large, the Poisson probability can be used.

The Poisson probability distribution is the foundation for a series of control charts for attributes. It can also be used as an approximation to the binomial probability distribution given the conditions that were discussed in this chapter.

SPOTLIGHT

In this and the following chapters, we spotlight elements of company wide quality management (CWQM). Probability and statistical process control are important CWQM techniques in the pursuit of customer satisfaction and competitiveness.

Japan has the Deming Prize named after an American statistician who was instrumental in establishing a national quality ethic in Japan. The United States did not have an equivalent award until recently, when the Malcolm Baldrige National Quality Award was instituted.

The purposes of the National Quality Award are to promote quality awareness, to recognize quality achievements of U.S. companies, and to publicize successful quality strategies.

Some companies apply for the award just to be audited. The award provides benchmarks against which a company can compare its progress in quality management with the U.S. best. Award winners are considered world class in their respective industries.

Companies that practice total quality management, sometimes called company-wide quality management, achieve high-quality performance, continuous improvement, and high levels of customer satisfaction, which translate into higher rates of financial return, than companies that do not believe in total quality.

1990 Examination Categories/Items

	Maximum Points
1.0 Leadership	100
1.1 Senior executive leadership	30
1.2 Quality values	20
1.3 Management for quality	30
1.4 Public responsibility	20
2.0 Information and Analysis	60
2.1 Scope and management of quality data and information	35
2.2 Analysis of quality data and information	25
3.0 Strategic Quality Planning	90
3.1 Strategic quality planning process	40
3.2 Quality leadership indicators in planning	25
3.3 Quality priorities	25
4.0 Human Resource Utilization	150
4.1 Human resource management	30
4.2 Employee involvement	40
4.3 Quality education and training	40
4.4 Employee recognition and performance measurement	20
4.5 Employee well-being and morale	20
5.0 Quality Assurance of Products and Services	150
5.1 Design and introduction of quality products and services	30
5.2 Process and quality control	25
5.3 Continuous improvement of processes, products, and services	25
5.4 Quality assessment	15
5.5 Documentation	10
5.6 Quality assurance, quality assessment, and quality improvement of support services and business processes	25
5.7 Quality assurance, quality assessment, and quality improvement of suppliers	20
6.0 Quality results	150
6.1 Quality of products and services	50
6.2 Comparison of quality results	35
6.3 Business process, operational, and support service quality improvement	35
6.4 Supplier quality improvement	30
7.0 Customer Satisfaction	300
7.1 Knowledge of customer requirements and expectations	50
7.2 Customer relationship management	30
7.3 Customer service standards	20
7.4 Commitment to customers	20
7.5 Complaint resolution for quality improvement	30
7.6 Customer satisfaction determination	50
7.7 Customer satisfaction results	50
7.8 Customer satisfaction comparison	50
Total points	**1000**

SUMMARY

- Probability statements deal with the likelihood of an event occurring. Probability is essential to the study of quality because it is the theoretical basis for sampling.
- Probability and statistics approach problems differently. In statistics, the quality analyst knows characteristics of a sample and infers information of the population. In probability, the quality analyst has information of the population and develops information about the sample pulled from the population.
- There are three approaches to probability: subjective, classical, and relative frequency. The subjective approach is based on the personal evaluation of the analyst to arrive at certain decisions. The classical approach is based on the assumption that outcomes, or events, of an experiment are equally likely to occur. The relative frequency approach assumes what occurred in the past will occur in the future.
- Two types of probabilities are discussed in this chapter: unconditional and joint probabilities. Unconditional probability is the simple probability of an event occurring. Joint probability is the probability of two or more independent events occurring in succession.
- Unconditional probability deals with independent events. It is calculated by summing the number of observations and dividing by the number of trials.
- Unconditional probability is used to calculate mutually exclusive and not mutually exclusive events. Events are mutually exclusive if one and only one event can occur at the same time. Two events are not mutually exclusive if both can occur at the same time.
- The probability of mutually exclusive events is the sum of the probabilities of the individual events.
- Venn diagram is a pictorial method for illustrat-

ing probabilities. The Venn diagram illustrating two mutually exclusive events is the sum of two noninterlocking circles.
- Elementary set theory can be used to express mutually exclusive and not mutually exclusive events.
- Joint probability of two events is the probability of each occurring in succession.
- Permutations and combinations are two probability techniques for calculating the number of groups and ungrouped objects. Specifically, a permutation is an ordered group of letters, objects, or products. A combination is an unordered group of letters, objects, or products.
- Three probability distributions are discussed in this chapter: binomial, hypergeometric, and Poisson. Each distribution must be used carefully.
- The binomial distribution describes the distribution of possible outcomes when only two events can occur, such as tossing a coin. It should be used when events are mutually exclusive, collected data are counted in whole numbers, the probability of the occurrence of each trial is the same, and the trials are independent of each other.
- The hypergeometric distribution describes the probability of accepting conforming products in samples taken without replacement from small populations.
- The Poisson distribution describes the probability of sequential events occurring, such as forecasting the probability rate of cars approaching a toll booth or patients going to an emergency care unit. The Poisson distribution assumes that the number of arrivals is a positive whole number, arrivals can be estimated from past history, the arrival rate is random, and the rate is constant during the period under examiantion.

KEY TERMS

binomial distribution   Probability distribution describing possible outcomes when only two events can occur.

classical probability   Approach to probability as-

suming outcomes of an experiment to be equally likely.

combination   Unordered group of letters, objects, or products.

event   Uncertain occurrence.

experiment   Process by which an observation or measurement is obtained.

factorial   Mathematical expression, represented by an exclamation point, used to abbreviate a series of descending numbers to be multiplied.

hypergeometric distribution   Probability distribution allowing the quality analyst to find the probability of acceptable products in samples taken without replacement.

joint probability   Probability of events, independent of each other, occurring successively; calculated by multiplying their individual probabilities.

mutually exclusive   One and only one event being able to occur at the same time.

permutation   Ordered group of letters, objects, or products.

Poisson distribution   Probability distribution allowing quality analyst to solve queuing problems.

probability   Statistical area of study where characteristics of the population are known and the quality analyst is concerned with calculating the likelihood of observing a desired sample.

probability distribution   Theoretical distribution showing how events are expected to vary under certain conditions and leading to inferences based on these conditions.

probability tree   Diagram illustrating the probability of successive independent events.

relative frequency   Approach to probability stating that the likelihood of a future event is the number of times the event occurred in the past divided by the total number of observations.

sample space   Set of all outcomes of an experiment.

statistically independent   Event independent of future events.

subjective probability   Approach to probability relying on the perspective of the analyst; similar to likelihood.

unconditional probability   Simple probability of an event occurring; calculated by dividing the number of observations by the number of trials.

Venn diagram   Interlocking or separate circles used to illustrate simple probability concepts.

QUESTIONS

1. Why is the study of probability important in the study of quality?

2. What is the difference between probability and statistics?

3. Discuss the differences among the three approaches to probability.

4. The relative frequency approach to probability is concerned with historical data and forecasting the future based on these data. What are the problems with this approach?

5. Give an example of a dependent and an independent event.

6. What is unconditional probability? Give an example.

7. What are mutually exclusive events? Illustrate two mutually exclusive events using a Venn diagram.

8. Give an example of a not mutually exclusive event.

9. What is the joint probability of having a head come up in three successive tosses of a coin? Construct a probability tree of this problem.

10. Calculate the number of letter permutations and combinations in the word BAR.

11. Calculate the number of combinations in the word DEGREE.

12. What is a binomial probability distribution?

13. What type of quality problems can the binomial probability distribution solve?

14. What is a hypergeometric distribution?

15. What type of quality problems can the hypergeometric probability distribution solve?

16. What is a Poisson distribution?

17. What type of quality problems can the Poisson distribution solve?

PROBLEMS

1. Find the probabilities of drawing the following cards from a deck of 52:
   a. A 6
   b. A red card
   c. A queen
   d. A red 6
   e. A 6 of diamonds

2. Find the probabilities of drawing the following cards from a deck of 52:
   a. A black card
   b. A king
   c. A black king
   d. A king of spades

3. Find the probability of drawing a king or a diamond from a deck of 52.

4. Find the probability of drawing a 10 or a red card.

5. What is the probability a couple's third child is:
   a. A girl, given the first child was a boy and the second a girl?
   b. A boy, given the first child was a girl and the second a girl?

6. In rolling a die, what is the probability of rolling a 4 four times in succession?

7. A bin has 25 parts, one of which is defective. The sampling table says for a lot of 25, pull two parts without replacement. What is the probability that one will be defective?

8. In Problem 7, what is the probability that two will be defective?

9. Using the word TRAIL, how many three-letter combinations can be made? Using the same word, how many three-letter permutations can be made?

10. A bin has 100 products and a sample of four is pulled without replacement. How many sample combinations can be made?

11. Calculate the following: C_0^5, C_0^{10}, C_0^{20}.

12. Calculate the following: C_1^7, C_1^{14}, C_1^{15}.

13. A continuous-production facility making polyethylene pipe has a fraction defective of 0.01. Inspection tables state that in samples of 20 products, if two or more defectives are found, then the sample is rejected. What is the probability of this occurring?

14. Referring to Problem 13, what is the probability if the fraction defective is 0.001?

15. Referring to Problem 13, what is the probability if only 10 products are sampled?

16. What is the probability of obtaining two defectives in a sample of two when the batch is 1% defective? Use the binomial distribution.

17. Referring to Problem 16, what is the probability of obtaining one defective in a sample of two? Again, use the binomial distribution.

18. A batch has 100 products and is known to have a fraction defective of 0.001. What is the probability that a sample of two will have one defective?

19. Assuming a Poisson distribution and $\lambda = 5$, calculate the following:
    a. $P(x = 1)$
    b. $P(x < 3)$
    c. $P(x = 5)$

20. Assuming a Poisson distribution and $\lambda = 3.2$, calculate the following:
    a. $P(x = 2)$
    b. $P(x < 3)$
    c. $P(x > 5)$

21. Assuming a binomial distribution with $n = 5$ and $p = 3$, calculate the following:
    a. $P(r = 3)$
    b. $P(r > 2)$
    c. $P(r > 4)$

22. Assuming a binomial distribution with $n = 10$ and $p = 0.05$, calculate the following:
    a. $P(r = 8)$
    b. $P(r < 3)$
    c. $P(r = 2)$

6

Attribute Control Charts

Attribute control **charts** summarize the output of a process, or operation, over time. **Attribute data** have only two values, such as good/bad, conforming/nonconforming, or acceptable/not acceptable. Attribute charts are used successfully in both manufacturing and service industries.

TYPES OF ATTRIBUTE CHARTS

What Are Attribute Data?

Attribute data are distinguished by their ability to be counted in terms of whole numbers, 1, 2, 3, ... For example, an operator can count the number of defects in a product or tabulate the number of errors in a standard form.

A product may have many measurable quality **characteristics.** A supervisor may not want to track the variation in a quality characteristic, but will want to know the number of defects in a product or the number of defectives in .an operation. She would use an attribute chart to track these data. Attribute charts are easy and quick to use, are cost-effective, and can identify areas needing attention and improvement.

Defects and Defectives

Attribute charts can be used to track defects as well as defectives. The term **defect** indicates a deviation from a specification in some quality characteristic of a product. A defect does not automatically cause a product to be rejected. It depends on the nature and severity of the defect.

The term **defective** describes the state of nonconformance of a product. A defective product may have one or more defects, each of which may differ in severity from critical to major to minor. One **critical defect** will automatically reject the entire product because a critical defect may jeopardize health and safety. A **major defect** will usually affect product function or performance. For example, a gear tooth not properly machined would affect function. A **minor defect,** such as a fabric tear, never seen by a customer, would not reject the product. A product can be rejected by having one critical or major defect or having several minor defects. Defects are also called nonconformities or discrepancies.

Illustration of an Attribute Chart

Management wants to know the daily and weekly defect rates of a product being produced. A daily attribute chart is used by operators and supervisors as the basis for monitoring current production and, if necessary, viewing the effects of any corrective action. A weekly attribute chart is prepared for management; it summarizes the quality levels of an area's operations. Shop foremen maintain daily charts, production supervisors maintain weekly charts.

In the plant each machine operator is a quality control inspector. There is no independent quality control group. The machine operator monitors his or her own production quality. At the end of the shift, each operator tells the shift supervisor how many defectives were produced. The supervisor tallies the numbers and plots the value on an attribute chart, which monitors the percentage of rejected items.

At the end of each day, shop supervisors for fabrication, assembly, and plating areas count the number of defective items out of the production run. The shift production run may be constant or vary. The chart for the work area is posted so everyone can see the current levels of quality. At the end of the week, supervisors tally the weekly totals and plot the value on an aggregate attribute chart.

Attribute charts can be used for many purposes, including:

- □ To summarize performance information.
- □ To establish a baseline level of quality, namely, average quality level or process average, for a machine, process, or work area. This baseline becomes the benchmark against which future improvements will be measured.
- □ To highlight improvements. Any improvement, meaning a trend line below the average level of quality, can easily be seen on the chart.
- □ To communicate information to management. As the old saying goes: ''a chart is worth 1000 words.'' Attribute charts convey much information very quickly.
- □ To highlight further investigation. Attribute charts are used to summarize performance information. They can pinpoint where an $\bar{X}$ and R chart

should be used, which is more sensitive to variation and can highlight specific assignable causes of variation.

PLANNING AND ANALYSIS

We discuss four different types of control charts:

- □ *p* **chart** for fraction defective of total products
- □ *np* **chart** for number of defective products
- □ *c* **chart** for number of defects
- □ *u* **chart** for number of defects per product/unit

The *p* and *np* charts are used to monitor defective products. The *c* and *u* charts are used to monitor defects. The *p* and *np* charts are based on the binomial distribution, the *c* and *u* charts on the Poisson distribution.

In this text the *p* chart is covered more comprehensively than other charts because the front-end planning and analysis of the *p* chart is the same for all attribute charts. Theory is kept to a minimum and is only introduced to enhance the discussion of the topic.

Any quality technique should only be used as long as it provides information useful for corrective action. Attribute charts are no exception and should be planned in a methodical fashion. The following steps are similar to the process steps discussed in the chapter on variable charts: define the purpose, define the process, identify management and operational requirements, and minimize variations.

Define Purpose. The purpose of the particular chart should be stated and understood by everyone in a work area. Attribute charts can display sensitive information and can be prominently posted to be scrutinized by anyone passing through the work area.

Attribute charts display more general information than variable control charts. One chart used at an inspection station or work area can serve to track different quality characteristics in each product or in a number of products. For example, in a large shop with different clusters of machines, one attribute chart can be used to track overall quality performance in terms of any number of factors, whether defective products, quality attribute defects, scrap costs, spoilage, or rework rates. Or one attribute chart may be constructed for each cluster of machines. For example, milling, grinding, boring, and cutting machines would each have a separate chart.

If even more detailed information is required, each separate machine may have an attribute chart tracking defective products or defects. Finally, to track dimensional variations in a specific machine, a variable chart would be used.

The number and type of charts are always related to the cost of their use. Some quality characteristics are critical and these would require a chart, regardless of cost.

Define Process. Attribute charts display overview information that should result in a better understanding and in some action to improve the process. For example, unusual conditions in one area may result in unusual variations, such as spikes, on the chart. Thus it is natural to investigate the problem where the chart is constructed. However, the problem may originate elsewhere. If it does, management should focus upstream to eliminate the root cause.

Identify Management and Operational Requirements. Product characteristics should be defined precisely if the relationship between the process and the desired output is to be understood. Standards, specifications, and drawings should be available for consultation. These identify critical, major, and minor quality characteristics. And, if problems or questions arise, an engineer or operations supervisor should be available for consultation.

Management and operational requirements and priorities should be articulated and understood by everyone so that attribute charts can be used properly. For example, if management mandates that scrap rates be reduced, it is counterproductive to chart and monitor labor absentee rates.

Minimize Variation. Assignable process or machine variations should be minimized before an attribute chart is used. This establishes a baseline from which any deviation can be identified. If an unusual event occurs, such as a lightning storm, training of a new operator, or even a plant walk-through by a manager, the attribute chart will show a spike or a trough, which would indicate process degradation or improvement.

p CHART

The *p* chart is an attribute chart that measures the proportion of defective products in a batch, lot, or shipment of products. This chart can also be used to measure the proportion of nonconforming defects.

The *p* chart is also called the **fraction defective** chart. The fraction defective is denoted by *p*. This is the ratio of defective products in one inspection, or successive inspections, divided by the total number of products inspected. This is usually expressed as a decimal. For example, if two products out of a sample of 25 are defective, the fraction defective is 0.08. The **percent defective** is the fraction defective number expressed as a percentage. So 0.08 expressed as a percent defective is 100 times the fraction defective, or 8.

When calculating process average and control limits, the fraction defective should be used. When *p* charts are displayed in a work area, the fraction defective is converted to percent defective to make it understandable to operating personnel.

Scan of Control Chart

The sample chart given in Figure 6.1 is used to graph all four attribute charts, namely the *p*, *np*, *c*, and *u* charts. It may be helpful to tour this attribute chart to become familiar with it. Although it is similar to the variable chart discussed in Chapter 2, this chart has notable differences. The attribute chart is only one

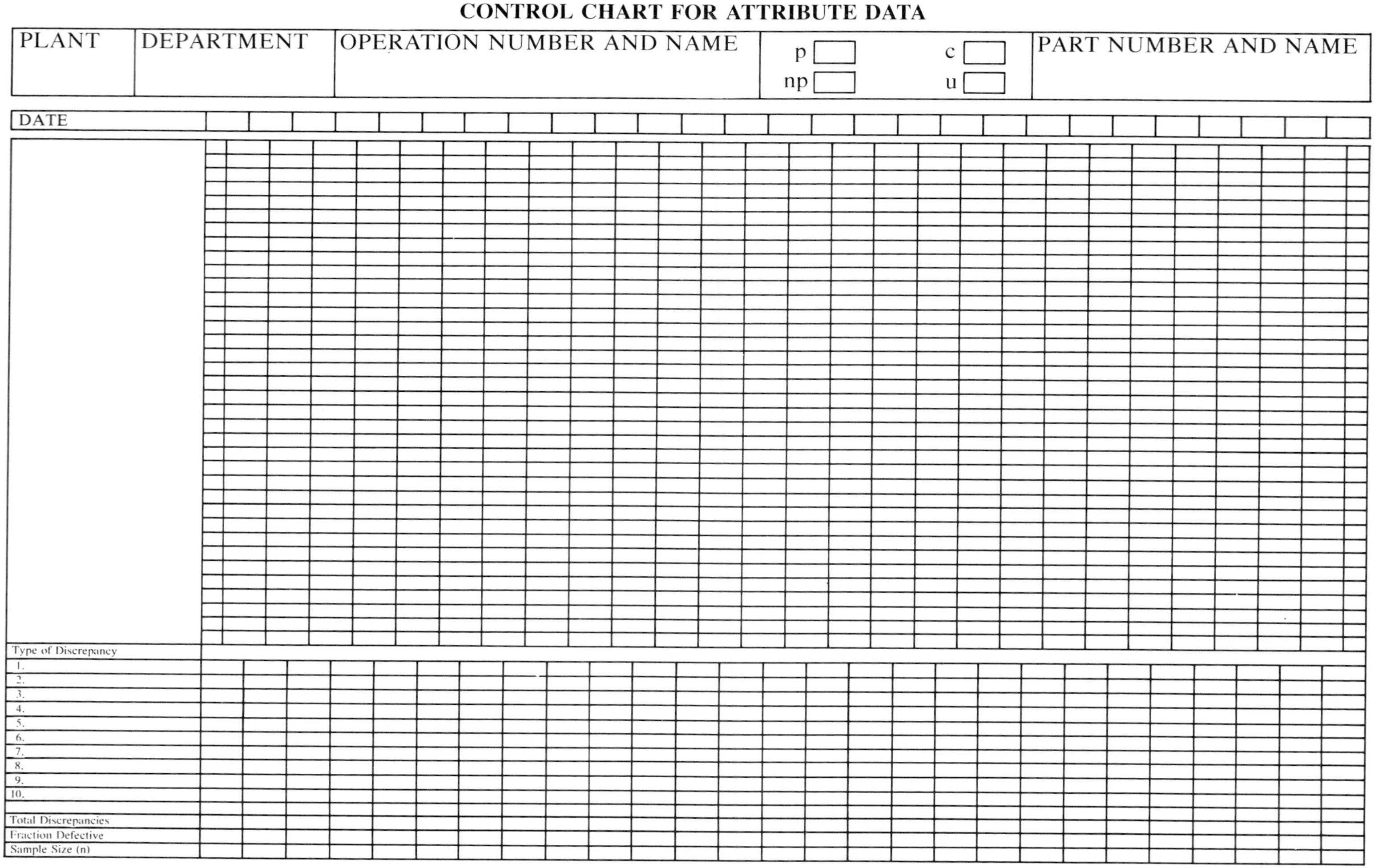

CONTROL CHART FOR ATTRIBUTE DATA

PLANT	DEPARTMENT	OPERATION NUMBER AND NAME	p ☐ np ☐	c ☐ u ☐	PART NUMBER AND NAME

DATE

Type of Discrepancy
1.
2.
3.
4.
5.
6.
7.
8.
9.
10.
Total Discrepancies
Fraction Defective
Sample Size (n)

FIGURE 6.1
Blank attribute chart.

graph. It is easier to construct than an $\overline{X}$ and R chart because only one set of control limits has to be calculated.

When the chart is first used, the operator fills out the top of the form. Typically the information may include part number, department, operation number and name, date, and type of chart.

On the bottom of the chart boxes list the type of discrepancy, total discrepancies, average or percentage of discrepancy and sample size. This particular chart can be used for all attribute charts to track the type of discrepancies.

When Should a *p* Chart Be Used?

P chart control limits are based on the binomial distribution. This distribution assumes that the probability of occurrence of each event is constant. In other words, the probability of a quality characteristic or product being rejected is constant from one unit to the next. Each product or quality characteristic being inspected is separate and independent from that of previous or succeeding products. *P* charts can be used to chart the following fraction defectives: parts coming from a screw machine, leaking diaphragms, group of quality characteristics from a wood table top.

A *p* chart could not be used to monitor products being coated by hot dipping in a chemical bath. If these products are pulled out of the bath too quickly, the coating will not have sufficient time to adhere properly. If 10 products are dipped at one time and removed too quickly, some if not all of the products will be flawed, and the probability of a product being accepted is therefore not constant. If one product is rejectable, then it is probable others are also.

P-Chart Construction

The basic steps for constructing a *p* chart apply equally to the other attribute charts. Only the formulas for calculating process limits and the process average are different. The following is the procedure for constructing a *p* chart:

- ☐ Select data
- ☐ Calculate fraction defective
- ☐ Calculate process average
- ☐ Calculate trial process limits
- ☐ Construct trial chart
- ☐ Maintain chart

Select Data. *P*-chart data must be accurate to reflect current operating conditions. Enough data should be collected to recognize shifts in conditions or performance. As a rule of thumb, the subgroup sample size should be large, consisting of 50 or more products. Also, the size of the subgroup can be constant or varying by no more than 25%.

The frequency of selection depends on the sensitivity of the chart to external disturbances. If the duration between selections is long, the chart will not display information from which the immediate cause of a deviation can be identified. If the duration between selections is too short, the chart will not show the trend line of any deviations. It may be necessary to track a quality characteristic over a long period of time to understand the nature and identify the trend line of the variation. The number of subgroups should coincide with the length of the data collection period and the amount of noticeable variations. It should at least cover 25 subgroup samples.

In our example we are tracking the discrepancies at the end of each shift in the assembly department of a furniture manufacturer. Five types of discrepancies (flaws) are monitored: chips, bruises, splits, looseness, and holes. Each shift, 50 pieces of furniture are produced and checked. This is our sample size n, even though it includes all products.

Calculate Fraction Defective. First, the subgroup data are recorded at the bottom of the chart. The number of defective products is counted in each sample subgroup. Then the fraction defective of each subgroup sample is calculated. The fraction defective is the number of defective products in a sample divided by the total number of products in that sample. Sometimes the fraction defective is called the proportion nonconforming. This is calculated using the following formula:

$$p = \frac{np}{n}$$

where p = fraction defective
np = number of defective products in subgroup
n = number of inspected products in subgroup

Calculate Process Average. Assuming the chart is being constructed for the first time, a minimum of 25 subgroups of data should be collected. The process average is calculated based on the subgroup data. It is the number of defective products in all of the subgroups divided by the total number of products in the subgroups.

The formula for the process average is

$$\bar{p} = \frac{np_1 + np_2 + \cdots + np_k}{n_1 + n_2 + \cdots + n_k}$$

where np_1 = number of defective products in first subgroup
n_1 = number of products in first subgroup
n_k = number of products in kth subgroup
$\bar{p}$ = process average

At the end of the first shift, one chip and one split in the wood were discovered. The fraction defective was

$$p = \frac{np}{n}$$
$$= \frac{2}{50} = 0.04$$

In our example this process average $\bar{p}$ is

$$\bar{p} = \frac{2 + 2 + 4 + \cdots + 1}{50 + 50 + 50 + \cdots + 50} = 0.0488$$

Calculate Trial Process Limits. First, trial process limits are calculated to determine whether the process is in control. If there are assignable causes, these are identified and eliminated. The upper control limit UCL_p and the lower control limit LCL_p are calculated using the process average. The control limits are the amount of variation that would be expected if the process or machine were in statistical control.

As with variable control charts, the upper control limit is plus three standard deviations above the process average while the lower control limit is minus three standard deviations below the process average. The formulas for UCL_p and LCL_p are as follows:

$$UCL_p = \bar{p} + \frac{3 \sqrt{\bar{p}(1 - \bar{p})}}{\sqrt{\bar{n}}}$$

$$LCL_p = \bar{p} - \frac{3 \sqrt{\bar{p}(1 - \bar{p})}}{\sqrt{\bar{n}}}$$

where $\bar{n}$ is the average sample size.

The calculated lower control limit can be a negative number if the sample size or fraction defective is small. If this occurs, the lower control limit is 0.

Control limits remain the same as long as the subgroup size stays constant. However, control limits will change if the subgroup size changes. Control limits have to be recalculated for each new subgroup size.

Using the preceding formulas for control limits, for $\bar{p} = 0.0488$ and $\bar{n} = 50$,

$$UCL_p = 0.0488 + \frac{3 \sqrt{0.0488(1 - 0.0488)}}{\sqrt{50}} = 0.140$$

$$LCL_p = 0$$

If subgroup sizes are not more or less than 25% of the average subgroup sample size, control limits can be calculated using the average subgroup sample size. If they do vary, then separate control limits have to be calculated for the periods with large or small samples.

Construct Trial Chart. A trial chart is constructed if it is not known whether the process is in or out of control. The chart is incremented according to the subgroup's fraction defective *p*. The vertical scale should be high enough to plot all of the data points. As a rule of thumb, this is 1.5 to 2 times the largest subgroup's fraction defective. Once scaled, each subgroup's fraction defective is plotted on the graph. The points are then connected (Figure 6.2).

Next the process average line is drawn as a solid line. The upper and lower control limits are drawn as dashed lines. All lines are appropriately labeled for easy recognition.

If unusual circumstances, situations, or conditions occur, these should be

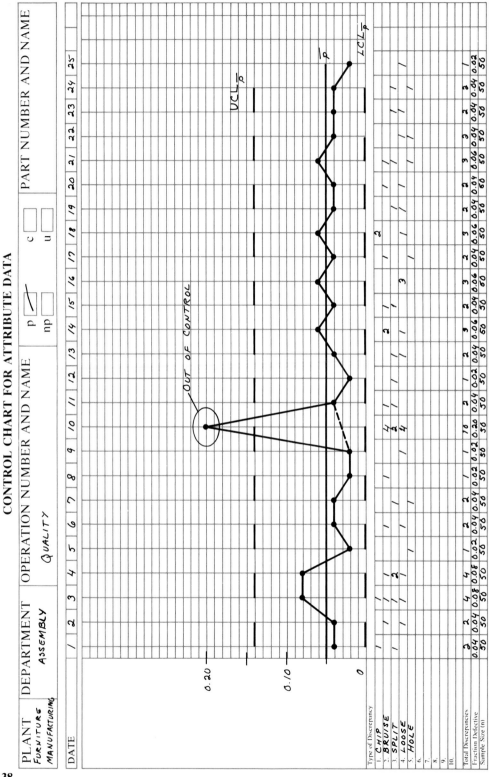

FIGURE 6.2
Sample *p* chart.

noted on a supplementary data sheet attached to the control chart. Sometimes results of present conditions may not appear on the chart until much later.

Maintain Chart. When the trial chart has been completed and trial limits have been computed, all points may be well inside the control limits, or some points may fall outside. If the chart is in control, the process average and the control limits should be left alone. If points are outside the control limits or if there is any abnormal variation, assignable causes should be identified, corrected, and eliminated. If many points are outside the control limits and causes cannot be identified, the process should be continued until an understanding evolves about the causes of variation.

In our example there is one out-of-control point, at the end of shift 10. The cause was discovered and eliminated. The process average is recalculated as follows:

$$\bar{p} = \frac{2 + 2 + 4 + \cdots + 1}{50 + 50 + 50 + \cdots + 50} = 0.0425$$

Note that the number of subgroups is now 24. Using $\bar{p} = 0.0425$ and $\bar{n} = 50$, the control limits are now

$$\mathrm{UCL}_p = 0.0425 + \frac{3\sqrt{0.0425(1 - 0.0425)}}{\sqrt{50}} = 0.128$$
$$\mathrm{LCL}_p = 0$$

The process is now in control and new control limits and process averages are drawn on the revised chart.

p-CHART INTERPRETATION

In attribute charts, a point above or below the control limits indicates process instability, a process out of control. Some assignable cause is influencing the process to create an unstable condition. Points above or below the control limits, or a nonrandom series of points, warrant immediate investigation.

Some management systems use a "management by exception" philosophy. As long as a system is in control, it is left alone. If a point goes above the upper control limit on the attribute chart, this signals the need for immediate attention and action. Otherwise management does not intrude into the daily operation of the process, relying on operators to maintain control.

If the process is in control, points below a positive nonzero lower control limit may suggest that operators have relaxed their inspection or lowered their standards; or, it might mean something is being done differently to improve the process drastically.

"Management by exception" is a static approach to management and quality—this can be very harmful. Changing customer needs require continuous improvement. Over time, a process in control will remain the same unless some fundamental changes occur to eliminate common causes (and fundamentally improve the system). This may be the redesign of a product, improved supplied material, additional training of personnel, or different production methods.

It should be mentioned that control limits will vary if the sample size varies by more than 25%. If the sample size does vary, the control limits have to be recalculated each time. This can become tedious. It is a problem for the operator or process engineer who has to use and interpret charts. If the operator knows that the sample limits have to be recalculated for each subgroup, he or she may pay more attention to the recalculation than to the process.

Out-of-Control Conditions

The following discussion applies to all four attribute control charts. The causes of out-of-control conditions are basically the same for the different charts. The following out-of-control conditions are discussed: out-of-control points and nonrandom patterns.

Out-of-Control Points. Out-of-control points are either above the upper control limit or below the lower control limit. Points above the upper limit usually indicate a worsening condition. A point may appear above the upper limit randomly or result from measurement error. Measurement error may be due to a calculation error or to an uncalibrated instrument. An out-of-control point may also reach the upper control limit by being part of an upward traveling trend (Figure 6.3).

Points below the lower control limit indicate the process has improved unusually well because of some assignable external cause. The cause should be determined immediately so that corrective action can be implemented and installed as part of the permanent system.

Nonrandom Patterns. Certain recognizable patterns of points inside the control limits may also indicate the existence of assignable causes. These patterns may be an early warning sign of future problems, or they may indicate a change in the level of performance during the period of the trend. Sometimes an operator has independently discovered something that improves the process. This is indicated by a gradual movement of points down the chart.

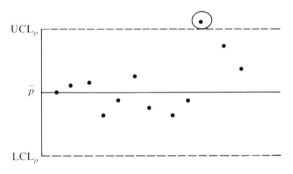

FIGURE 6.3
Example of an out-of-control point.

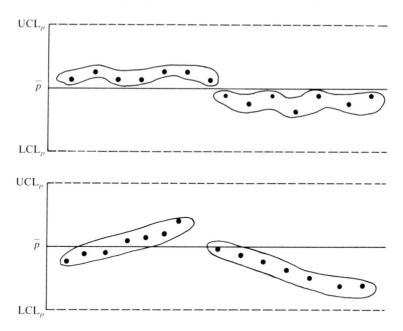

FIGURE 6.4
Example of out-of-control conditions.

If a process is in control, approximately an equal number of points should fall on either side of the process average. But unnatural patterns or runs can occur. Specifically, out-of-control conditions exist if seven points in a row are above or below the process average or if seven points in a row are continuously rising or falling (Figure 6.4).

np CHART

The p and np charts are very similar. The p chart graphs the fraction defective. The np chart displays the actual number of nonconforming products. The number of nonconforming, or defective, products is calculated by multiplying the number of defectives p in the sample times the size n of the sample (Figure 6.5).

Both p and np charts can be used to display the same data. The np chart is preferred if the size of the subgroup sample is constant throughout the chart. The chart does not require calculation of the percentage of nonconforming products in each subgroup sample.

Method for Constructing the *np* Chart

The method for constructing the np chart differs from that of the p chart in the following ways: the sample size must be constant, the scales graph different information, and the control limits are calculated differently.

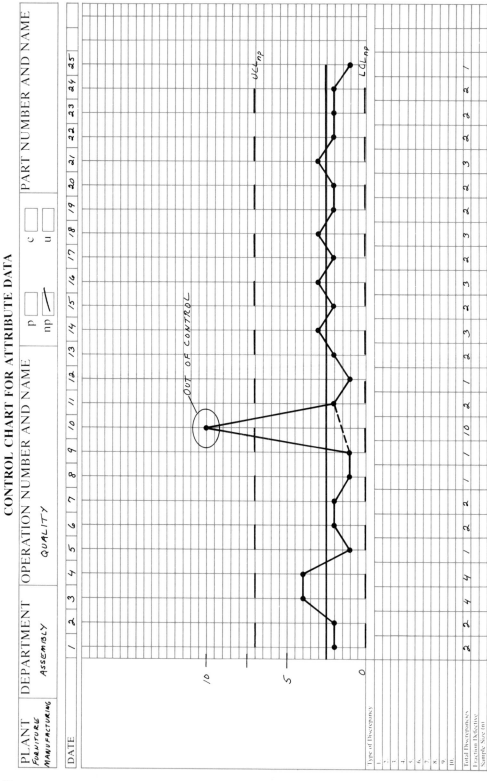

FIGURE 6.5
Sample *np* chart.

Sample Size Is Constant. If the subgroup size varies, *p* and *np* charts vary. In the *p* chart, as the sample group size varies, the process average remains the same and process limits change for each subgroup. In the *np* chart, if the sample size remains constant, neither the process average nor the process limits change.

The frequency of sampling is determined by the operating process and by management requirements. As mentioned, the frequency of sampling should be adjusted so that variations can be detected. Sometimes this is a trial-and-error process. If the product is critical to safety, the frequency of sampling may be increased to ensure that quality levels are maintained.

Scales Graph Different Information. The vertical scales on the *np* chart are also different from those of the *p* chart. On the *np* chart the vertical scale tracks the actual number of defective products, while it tracks the fraction of rejected parts on the *p* chart.

The process average of the *np* chart is similar to its *p*-chart counterpart. However, the calculation is different. The *np* process average is calculated using the following formula:

$$n\bar{p} = \frac{np_1 + np_2 + \cdots + np_k}{k}$$

where $n\bar{p}$ = defective process average
np_1 = number defective in first subgroup sample
k = number of subgroups

Using the data from our furniture manufacturing example, the process average $n\bar{p}$ is calculated as follows:

$$n\bar{p} = \frac{2 + 2 + 4 + \cdots + 1}{25} = 2.44$$

Control Limits Are Calculated Differently. Control limits on the *np* chart are calculated differently. The formulas for the upper control limit UCL_{np} and the lower control limit LCL_{np} are:

$$UCL_{np} = n\bar{p} + 3\sqrt{n\bar{p}\left(1 - \frac{n\bar{p}}{n}\right)}$$

$$LCL_{np} = n\bar{p} - 3\sqrt{n\bar{p}\left(1 - \frac{n\bar{p}}{n}\right)}$$

where $n\bar{p}$ = process average nonconforming or defective
n = subgroup sample size

Using the preceding formulas, if $n\bar{p} = 2.44$ and $n = 50$, the trial control limits are

$$UCL_{np} = 2.44 + 3 \sqrt{2.44\left(1 - \frac{2.44}{50}\right)}$$

$$= 2.44 + 3\sqrt{2.3209} = 7.01$$

$$LCL_{np} = 2.44 - 3\sqrt{2.3209} = 0$$

As in the p chart, if there are out-of-control points or nonrandom patterns, these have to be eliminated before a process can be considered in control. The control limits are recalculated based on the remaining in-control points. For the 24 subgroups, with $n\bar{p} = 2.125$ and $n = 50$,

$$UCL_{np} = 2.125 + 3\sqrt{2.125\left(1 - \frac{2.125}{50}\right)}$$

$$= 2.125 + 3\sqrt{2.035} = 6.40$$

$$LCL_{np} = 0$$

c CHART

The c chart graphs the number of defects in a product or discrepancies (errors) in a form. In the chart shown in Figure 6.6 we track the total number of defects in the body casting of each valve. Notice that we track the number of defects, not the type of defect. The c chart requires a constant sample size. Each subgroup in the c chart can represent a single product, or it can be the number of defects in a single product.

The subgroup sample size should be constant, so among different subgroups there is an equal probability for defects to occur. If the probability of defects changes from subgroup to subgroup, the u chart for defects per unit should be used.

C charts are based on the Poisson distribution because the chances of a defect occurring are numerous, but the chance of a defect occurring in any subgroup at any point in time is negligible.

C charts can be used to:

☐ Count the number of holidays (scratches) on a painted fender
☐ Count the number of flaws in a bolt of material
☐ Count the number of defects on a truck
☐ Count the number of defects in a gear assembly

Method for Constructing the *c* Chart

The method for constructing a c chart is similar to that for other control charts, but it is different in the calculation of the process average and the control limits. The process average is calculated using the following formula:

$$\bar{c} = \frac{c_1 + c_2 + \cdots + c_k}{k}$$

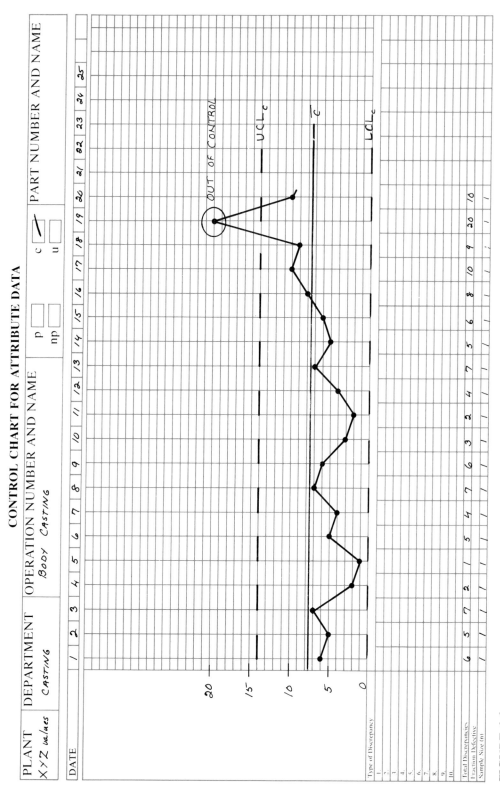

FIGURE 6.6
Sample *c* chart.

145

where $\bar{c}$ = process average
k = number of sample subgroups
c_1 = number of defects in each sample subgroup

The process average for our furniture example is calculated as

$$\bar{c} = \frac{6 + 5 + 7 + \cdots + 10}{20} = 6.35$$

The upper control limit UCL$_c$ and the lower control limit LCL$_c$ formulas are:

$$UCL_c = \bar{c} + 3\sqrt{\bar{c}}$$
$$LCL_c = \bar{c} + 3\sqrt{\bar{c}}$$

where $\bar{c}$ is the process average.
For $\bar{c} = 6.35$,

$$UCL_c = 6.35 + 3\sqrt{6.35}$$
$$= 6.35 + 7.56 = 13.91$$
$$LCL_c = 6.35 - 7.56 = 0$$

Interpretation

The c chart is interpreted similarly to the other attribute charts. Again, if one or more points are above or below the control limits, these are investigated and eliminated. Once eliminated, the process average and the control limits are recalculated without the out-of-control points. The new process average and control limits serve as the basis of a process in control.

In our example we have one out-of-control point. We thus recalculate the process average and the upper and lower control limits based upon the remaining 19 subgroups:

$$\bar{c} = \frac{6 + 5 + 7 + \cdots + 10}{19} = 5.63$$
$$UCL_c = \bar{c} + 3\sqrt{\bar{c}}$$
$$= 5.63 + 3\sqrt{5.63} = 12.75$$
$$LCL_c = \bar{c} + 3\sqrt{\bar{c}}$$
$$= 5.63 - 7.12 = 0$$

All of the remaining points are within the revised control limits.

u CHART

The u attribute chart is developed in a similar way as the c chart. Also, the u chart can be used to graph the same attributes as the c chart. At least 20 subgroup samples are gathered, and the trial control limits are calculated. If there are out-of-control conditions, these are investigated and the causes eliminated. Then the revised limits are calculated.

The *u* chart measures the number of defects per product. It is similar to the *c* chart, except that the number of defects are expressed on a per unit basis. The sample subgroups can vary in size (Figure 6.7). In our example, we measured the defects per truck at the end of a day's production.

u-chart calculations follow the pattern of previous attribute charts. First defects per unit in each sample subgroup are calculated:

$$u = \frac{c}{n}$$

where u = defects per unit
$\quad c$ = number of defects discovered in sample subgroup
$\quad n$ = sample subgroup size

In our example the number u, defects per unit in the first sample subgroup is

$$u = \frac{c}{n} = \frac{2}{8} = 0.25$$

The formulas for the upper control limit UCL_u and the lower control limit LCL_u are

$$\text{UCL}_u = \bar{u} + \frac{3\sqrt{\bar{u}}}{\sqrt{\bar{n}}}$$

$$\text{LCL}_u = \bar{u} - \frac{3\sqrt{\bar{u}}}{\sqrt{\bar{n}}}$$

where $\bar{n}$ = average sample subgroup size
$\quad \bar{u}$ = process average

In our example,

$$\bar{u} = \frac{0.25 + 0.33 + \cdots + 0.375}{20} = 0.39$$

$$\bar{n} = \frac{8 + 9 + 7 + \cdots + 8}{20} = 7.95$$

Using the preceding formulas for $\bar{u} = 0.39$ and $\bar{n} = 7.95$,

$$\text{UCL}_u = 0.39 + \frac{3\sqrt{0.39}}{\sqrt{7.95}}$$

$$= 0.39 + \frac{3 \times 0.624}{2.82} = 1.05$$

$$\text{LCL}_u = 0.39 - \frac{3\sqrt{0.39}}{\sqrt{7.95}}$$

$$= 0.39 - \frac{3 \times 0.624}{2.82} = 0.0$$

If LCL_u is negative, the lower limit becomes 0.

Since this process is in control, that is, all plotted points are within the control limits, revised limits do not have to be calculated.

Interpretation of the *u* chart is similar to that of the *p* chart.

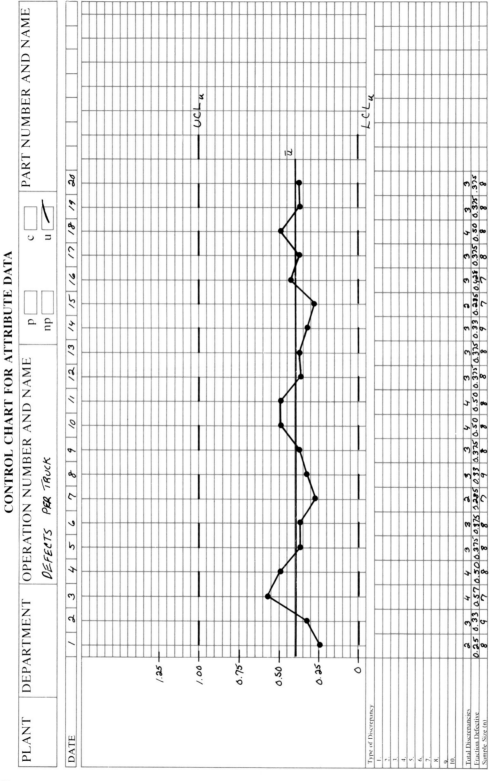

FIGURE 6.7
Sample *u* chart.

SPOTLIGHT

Westinghouse Electric Corporation's Commercial Nuclear Fuel Division (CNFD) was a 1988 Malcolm Baldrige National Quality Award winner. Its story reflects what is necessary to become a winner of this award using Company Wide Quality Management, and process control to increase safety and satisfy customers.

CNFD engineers, manufactures, and supplies pressure water reactor fuel assemblies for commercial nuclear power reactors. The fuel in these assemblies generates heat through nuclear fission, which is converted into electricity.

CNFD supplies about 40% of the U.S. light-water reactor fuel market and 20% of the free-world market. CNFD attributes its leadership to the quality of its products and services.

Commercial nuclear power reactor fuel is bundled into assemblies of enriched uranium fuel rods. A typical electric nuclear power plant of 1100 megawatts has a core of about 193 fuel assemblies, consisting of 51,000 fuel rods containing over 18 million uranium dioxide fuel pellets and over 700,000 feet of metal tubing.

Each fuel rod and all of its fuel pellets must perform at maximum efficiency and reliability for up to 5 years. The temperature at the center of the pellets exceeds 1700°F, and the tubing must sustain pressures of more than 2200 psi.

Under these conditions quality standards are high. If a failure occurs, the plant must shut down, which is expensive, and repairs are hazardous. The reliability of Westinghouse's nuclear fuel assemblies therefore approaches zero defects.

Early Quality Program

Federal and industry regulations govern the design, construction, and operation of commercial nuclear power plants. The regulations stress paper work and womb-to-tomb documentation of material traceability, product inspection, procedures, and continuous customer auditing.

This type of compliance is largely restrictive. Something is done, and paperwork is generated showing how, when, and why it was done by a particular person. The paperwork can later be audited by a regulatory body. The purpose of the womb-to-tomb audit trail is to provide regulatory auditors a level of assurance that quality is being pursued.

In the early 1980s the marketplace changed dramatically. New nuclear plant construction stopped. CNFD saw a market shift from providing fuel to new plants to providing new fuel for existing plants. At the same time, foreign competition increased.

Value Leadership

External competition necessitated changes in accepted procedures and management practices. Westinghouse realized that leadership through long-term quality or, in other words, creating value was the means to compete successfully. Management understood that its primary mission was to create value for the organization's stakeholders, including customers, stockholders, employees, and the community. Westinghouse believed that the concept of total quality was the fundamental means, or strategy, for creating value.

Westinghouse defines total quality as "performance leadership in meeting customer requirements by doing the right things right the first time." Hopefully the results are high-quality products and services that satisfy the customer. To accomplish this, Westinghouse also believed the obsession for quality must pervade every element of the organization, from product design to development, sourcing, production, human resource management, and distribution.

As a result, CNFD did some soul searching, trying to recognize what it did well and to find a competitive market edge. CNFD found the answer in continuous, long-term quality improvement. So each year, starting in 1983, CNFD developed quality plans and achievable targets in its continuous improvement journey.

The following yearly targets illustrate the journey to total quality:

1983   Create and increase general employee quality awareness.
1984   Track and report quality failure costs.
1985   Establish CNFD measurement indicators to monitor progress.
1986   Create mission statement and second- and third-tier measurement indicators.
1987   Focus on the external customer.
1988   Focus on total quality throughout the organization.

Four Imperatives

The CNFD model is built on the four Westinghouse total quality imperatives: quality management systems, products, technology, and people, which, it is hoped, will result in customer satisfaction.

Customer Satisfaction

CNFD's primary customer is the end user, who is the owner and operator of commercial nuclear power plants. These plants are owned by public utilities. Another important customer is the federal government and its various regulatory arms, including the Nuclear Regulatory Commission.

Commercial nuclear power reactors must be safe and reliable. If a nuclear power plant cannot be controlled while under operation, the results can be catastrophic. Public safety is paramount, as was illustrated in the Chernobyl and Three Mile Island disasters. Also, if a plant has to be repaired, it must be done quickly, effectively, and safely. Downtime on a multibillion dollar plant is expensive.

Quality Management Systems

Total quality must be all pervasive in the daily operating strategy for organizations that design, construct, operate, maintain, and supply these plants. Top management in these organizations must commit itself and decide that it wants to be a world-class organization.

To become the best, CNFD realized that long-term commitment was necessary to transform its working environment from filling out forms and maintaining a documentation trail to an organization that was quality obsessed and customer driven.

CNFD's quality council initiated and directed the quality process. The quality council was comprised of the people who manage operations. CNFD did not put a single person in charge of quality because this would send the wrong message to people. CNFD wanted to send the message that everyone was responsible for the quality of his or her efforts.

Quality People

CNFD believes that customer satisfaction can only be attained through human resource excellence. Management first establishes an organizational climate that permits employees to take personal accountability for quality and then empowers them with the authority to make positive changes to improve quality continuously.

Creating a quality culture is essential if a quality ethic is to become ingrained in the organization. Culture is the pervasive attitude or ethos that permeates the organization. In a participative management culture, employees are empowered with the resources and authority to meet customer requirements by doing the right things right the first time.

The goal of continuous individual improvement is at least as important as product or process improvement. To promote a company-wide quality culture, employees are encouraged to participate in quality-related training programs.

Suggestions are encouraged and employee achievements are sincerely recognized and rewarded. At CNFD, employees increased the number of suggestions from 25 to 2000 submittals in a 3-year period.

Quality Products and Technology

To attain higher quality objectives and pursue continuous improvement, CNFD incorporates the latest engineering and manufacturing technologies, including statistical process control (SPC), computer-aided design (CAD), robotics, and just-in-time (JIT) manufacturing.

Statistical process control systems helped CNFD limit variation by allowing operators to fine-tune processes to keep products closer to the specification targets. Thus the operator, not the quality control inspector or analyst, is responsible for quality. One team at CNFD's specialty metals plant was notably successful in reducing the wall thickness variation of nuclear tubing by in-process statistical monitoring and adjustment.

CNFD uses just-in-time concepts in many operations. As the name implies, products arrive just in time to be used by the process operator. The time it takes to perform a process or task is reduced by eliminating unnecessary delays, rework, and scrap. Quality is improved and cycle time, the time it takes to perform all necessary operations, is reduced.

The fundamental measure of CNFD's quality is fuel assembly reliability in the commercial nuclear reactor. Reliability is measured by monitoring the amount of radioactivity that leaks from defective fuel rods in the coolant water of the reactor system. CNFD's goal is zero defective fuel rods.

To achieve this quality level, information flows continuously throughout the organization, focusing on improving product quality and service performance. In addition to product reliability, on-time deliveries and first-time yields of manufactured components were measured.

Results

Total quality has been the major factor in CNFD's overall business strategy and success. As a result of continuous quality efforts, CNFD's primary quality measure, fuel reliability as measured by the number of leaking rods, has improved by a factor of 10 in a few years. Out of 2.5 million fuel rods in operation today, less than 100 are known or presumed to be leaking. Fuel reliability performance has attained world-class performance standards and approaches 99.9995%.

How does this translate to customer satisfaction? CNFD has been able to increase its customer satisfaction rating by 6% in less than 1 year. This means customers attach greater value to CNFD's products and services than they did a year ago. This translates to new and repeat business. CNFD is enjoying record new orders and has an extensive backlog.*

It seems that quality does pay off.

---

* Adapted from the Westinghouse study "Performance Leadership Through Total Quality," submitted to National Institute for Standards and Technology, 1989.

SUMMARY

- Attribute charts summarize the output of a process over time in terms of acceptable/not acceptable, go/no-go, and conforming/nonconforming. Attribute charts are used in manufacturing and service organizations.
- The terms defect and defective are often used. Defect is a deviation from a specification in a quality characteristic of a product. Defective describes the state of nonconformance of the entire product.
- Defects are rated in terms of three levels: critical, major, and minor. A critical defect, such as a failed valve in a mechanical heart, causes a product to malfunction with a possible loss of life. A major defect, such as a faulty resistor in a radio, causes a product to malfunction. A minor defect, such as a surface blemish, is more of a nuisance.
- This chapter introduces four types of attribute charts: p, np, c, and u charts. The p chart measures fraction defective products. The np chart measures the number of defective products. The c chart measures the number of defects. The u chart measures the number of defects per unit.
- Attribute charts, like variable charts, should not be used haphazardly; careful planning is essential. Otherwise data will be generated that do not improve the process. The steps for planning the use of attribute charts can be divided into the following: define the purpose of the chart, define the process, identify management and operational requirements, and minimize variation.
- The p chart measures the proportion of defective products in a batch. The fraction defective is the decimal equivalent of the fraction of products that are defective in a shipment. For example, 10 products out of a sample of 100 are defective. The fraction defective is 0.10. The percent defective converts this number to a percentage; for example, 0.10 becomes 10%.
- The p chart is similar to the other charts in terms of planning and construction. Calculations of the control limits are different. The p chart is based on the binomial distribution. The probability of a quality characteristic being rejected is constant from product to product.
- The steps for constructing the attribute charts are similar, specifically: select data, calculate process average, calculate trial process limits, construct trial chart, and maintain chart.
- Attribute charts also have upper and lower control limits that represent plus and minus 3σ, respectively. The purpose of attribute chart limits is to warn of aberrations in the process.
- Two general out-of-control situations are out-of-control points and nonrandom patterns. If points go above the upper control limit, an assignable factor is causing the deviation. If points go below the lower control limit, there is a faulty measurement or an improvement in the process.
- Nonrandom patterns exist if seven points in a row are above or below the process average, or if seven points in a row are continuously rising or falling.
- The np chart displays the actual number of nonconforming products instead of the fraction defective on the p chart. The np and p charts can display the same data.
- The c chart requires a constant sample size and graphs the number of defects in a product or form. The letter c represents the count of defects in a single product.
- The u chart graphs the number of defects per product. The sample subgroup size can vary.

KEY TERMS

attribute data   Quality data that can be counted.

attribute chart   Graph measuring quality in terms of the presence of attributes, or characteristics, in a sample.

c chart   Attribute chart monitoring the number of product defects in a sample.

characteristic   Quality-related attribute of a product.

critical defect   Defect causing automatic rejection of the entire product.

defect   Deviation from a specification in some product quality characteristic.

defective State of nonconformance in a product.

fraction defective Ratio of defective products in one inspection, or successive inspections, divided by the total number of products.

major defect Defect affecting product function or performance.

minor defect Cosmetic defect, not affecting performance.

np **chart** Attribute chart for evaluating a process in terms of the total number of defective units in a sample.

p **chart** Attribute chart for evaluating a process in terms of the fraction defective of the total number of units in a sample.

percent defective Percentage of products in a lot or batch that are defective.

u **chart** Attribute chart monitoring the number of defects per product in a sample.

QUESTIONS

1. Where are attribute charts used?

2. What is the difference between defects and defectives? Give an example of each.

3. What are three levels of defects? Give an example of each.

4. How can attribute and variable charts be used together?

5. *P* and *np* charts are based on the binomial distribution. This means the charts should be used only in certain situations. Explain.

6. *C* and *u* charts are based on the Poisson distribution. This means the charts should be used only in certain situations. Explain.

7. Explain the planning methodology prior to using attribute charts.

8. Why should management requirements be explained before implementing attribute charts?

9. What are the relative advantages and disadvantages of each of the attribute charts? When should each be used?

10. What is the difference between fraction defective and percent defective? Give an example of each.

11. When should the control limits of an attribute chart be recalculated?

12. What does "management by exception" mean?

13. When points are above the upper control limit or below the lower control limit, what does it mean? What should be done upon discovering this condition?

14. Give an example of a nonrandom pattern. Explain what should be done upon discovering this condition.

15. What charts should be used if the sample size differs?

16. Give an example where the *c* chart can be used.

PROBLEMS

1. Construct a *p* chart for the following data. The subgroup size is constant, 100 products per shift. Interpret the chart.

Shift	Number of Products Checked	Number of Defective Products	Shift	Number of Products Checked	Number of Defective Products
1	100	4	11	100	3
2	100	5	12	100	4
3	100	3	13	100	4
4	100	4	14	100	3
5	100	5	15	100	3
6	100	5	16	100	2
7	100	4	17	100	4
8	100	3	18	100	0
9	100	4	19	100	4
10	100	8	20	100	3

2. Construct a p chart for the following data. The subgroup size changes per shift. Interpret the chart.

Shift	Number of Products Checked	Number of Defective Products	Shift	Number of Products Checked	Number of Defective Products
1	105	4	11	95	3
2	110	5	12	110	4
3	102	3	13	83	4
4	115	4	14	98	3
5	90	5	15	95	3
6	86	5	16	85	2
7	101	4	17	100	4
8	99	3	18	101	0
9	109	4	19	88	4
10	90	8	20	94	3

3. Using the data shown below, determine the centerline, the control limits, and draw a p chart.

Subgroup	Number Inspected	Number Defective	Subgroup	Number Inspected	Number Defective
1	200	4	9	200	7
2	200	8	10	200	7
3	200	2	11	200	6
4	200	5	12	200	4
5	200	7	13	200	1
6	200	8	14	200	15
7	200	9	15	200	5
8	200	5			

4. The count of tears and rips in each bolt of material is given in the table below. Construct a c chart determining the trial centerline and control limits; then calculate the revised limits when the out-of-control points have been eliminated.

Code	Defect Count	Code	Defect Count
1	5	11	6
2	6	12	7
3	6	13	5
4	4	14	6
5	6	15	4
6	5	16	5
7	0	17	5
8	5	18	6
9	6	19	10
10	5	20	12

5. Calculate the trial and the revised control limits using the data in the table shown below. The table totals the defects of each shipment of five products from ACME Production.

Shipment	Size	Total Defects	Shipment	Size	Total Defects
1	5	12	13	5	10
2	5	13	14	5	11
3	5	11	15	5	11
4	5	12	16	5	13
5	5	11	17	5	10
6	5	9	18	5	13
7	5	10	19	5	9
8	5	6	20	5	10
9	5	9	21	5	18
10	5	9	22	5	12
11	5	11	23	5	11
12	5	12			

6. Control limits are set for the diameter of a bar. The population fraction defective is 0.01. Find the control limits for 500 products a day, for 2000 products a day, and draw the p chart.

7. An np chart is used in a plating process which is in statistical control. Every 8-hour shift 50 parts are inspected. The population fraction defective is 0.05. Calculate the control limits and the centerline.

8. A sample of 100 items is pulled from a process in control. The fraction defective is 0.05.
   a. What is the expected average number of defective units per sample?
   b. Calculate the control limits of an np chart.

9. The process average of a machine is 1% defective product. Each day 1000 units are produced, 50 products are sampled, and the data are plotted on an np chart. Calculate the process limits of this process.

10. Assembly-line employees attach parts and assemblies onto a truck chassis. The complete chassis is checked and all defects are repaired. The number of defects per chassis for the first 15 trucks are:

Chassis	Number of Defects	Chassis	Number of Defects
1	0	9	1
2	1	10	2
3	0	11	1
4	0	12	6
5	2	13	2
6	3	14	1
7	1	15	1
8	1		

a. Design a c chart and plot and analyze the data.

b. If there is an out-of-control condition, recalculate the control limits.

7

Inspection and Sampling

While this text emphasizes prevention and continuous improvement, sampling and inspection are still used in certain applications. This chapter focuses on planning and managing inspection and sampling instead of covering many of the technical details of inspection which have limited usage in company-wide quality management.

Inspection is the actual checking of quality attributes for conformance to specifications. **Sampling** is a probability-based technique that, given certain assumptions, indicates the number of products to be pulled from a lot, or batch, for inspection. Sampling may be 100% or it may be a small percentage of the population. Sampling tables have been developed that specify the number of products to be selected from a population. The number of products pulled is determined by the risk to the consumer, the risk to the manufacturer, cost, and the **acceptable quality level.**

WHAT IS INSPECTION?

Inspection is used in the inspection and prevention modes of quality management. In traditional inspection, a quality control inspector measured product characteristics after a production operation. In the prevention mode, an operator samples products from his or her production and self-inspects the output of his or her operation. A quality control inspector no longer catches defective products after an operation. For example, in statistical process control (SPC), the operator periodically pulls four or five products from production and measures a dimensional characteristic. The operator is an inspector and manager of the process.

Inspection is still used in manufacturing and service operations to check quality. Manufacturing operations process large numbers of products, and service operations process large volumes of paperwork. Automatic test equipment permits 100% testing of certain products. For example, inspection tollgates in production facilities are established to determine the quality level at set points on the production line before sending products to the next operation.

In manufacturing, inspection may occur in three major operational areas: on incoming material, in process, and as final inspection once production is completed.

Incoming Material Inspection. **Incoming material inspection** may be a visual or a dimensional check of quality attributes of incoming material from external or internal suppliers. Incoming material may be raw material, partially finished goods, parts, or capital equipment.

Incoming material is usually checked in a dedicated inspection area next to the loading dock. In this area, different measuring instruments, such as destructive, nondestructive, and computer-controlled equipment, are able to measure a product dimension to millionths of an inch. This area is sometimes a special temperature-, dust-, and humidity-controlled area.

In-Process Inspection. **In-process inspection** evaluates the quality of products being machined or processed. In a manufacturing plant, products move from one operation to another in a continuous batch, or some combination mode. As a batch of products moves from one area to another, the batch, or lot, is inspected to ensure that previous operations manufactured products properly. For example, as raw aluminum sheets arrive on the loading dock, they go directly to the shearing area to be cut to specified dimensions. Then they go to the forming area to be shaped and later to the drilling area, where holes are punched in a preset pattern. Finally parts travel to the paint booth to be blasted, primed, and painted. At the beginning of each of these production operations, inspection tollgates inspect the quality of the previous operation.

Final Inspection. By the time material arrives at **final inspection**, much work, or value, has been added. So final inspection is a test of the overall product's performance, appearance, reliability, or serviceability.

Final inspection usually does not measure internal dimensions. It is assumed that critical measurements and tolerances were maintained while in process. Also by this time, the dimensions of internal components cannot be checked because they are inaccessible.

Disposition of Products

Following any inspection, a batch of products is **accepted** or rejected. If a batch is accepted, it is sent directly to the next production step or placed in inventory. If placed in inventory, products are pulled out as needed.

As just-in-time (JIT) production is adopted by more companies, products are sent directly onto the production line. As the name implies, products from suppliers are not inspected and are assumed to be 100% defect-free. With this assumption there comes the risk that a shipment of defective products can shut down a JIT production line.

Lots, or batches, are sometimes accepted for use after having failed inspection if demand requires immediate use and if nonconformances do not affect customer satisfaction, safety, or performance. Engineering or a material review board must authorize the deviation from a specification. The deviation authorization should be a one-time effort. The supplier or the source of the nonconforming material is still required to supply conforming products in subsequent shipments.

Rejected Products. Rejected products may be returned, reworked, or scrapped. When a shipment of products is rejected, products are returned to the source. If the source is a supplier, the supplier may check 100% of the products in the shipment and separate the acceptable from the unacceptable. If a shipment is returned, the supplier is expected to assume shipping and inspection costs, provide conforming products, and improve future product quality.

Rejected products can also be sent to the source for **rework.** Rework assumes that the defect can be remedied in a cost-effective fashion. In some cases this is impossible. For example, if a machine operator drilled holes in wrong locations, the product will be scrapped. This is especially expensive if products have gone through many production steps that have added considerable value.

Rejected products can also be scrapped. This is expensive and should be avoided. Prevention hopefully eliminates the option of scrapping products. However, if products are scrapped, disposal should be safe and comply with environmental regulations. Many products are covered by federal and state environmental statutes that regulate the storage, transport, and disposal of these products.

INSPECTION PLANNING

Any inspection must be planned before being implemented. Proper planning ensures that scarce organizational resources are focused on solving high-priority problems. Any assignable cause variation in personnel, techniques, methods, procedures, or equipment should be minimized before inspection. As we have stressed throughout this text, unnatural variation is the cause of poor quality.

Inspection planning is a series of steps that include determining inspection requirements, preparing for inspection, and defining methods. This planning process can be used in manufacturing, service, and construction activities. Specifically, the quality analyst must:

- □ Identify customer requirements
- □ Identify managerial responsibilities
- □ Identify required resources
- □ Identify quality attributes
- □ Determine the frequency of inspection

☐ Define inspection methods and acceptance criteria
☐ Report inspection results

Identify Customer Requirements. The first step in any inspection is to identify the internal customer and his or her requirements. Inspection is costly and its purpose should be justified in terms of satisfying internal customer requirements.

Once the purpose is determined, its scope is formulated. The scope is the breadth of the inspection. The inspection scope must be identified carefully so that it does not become an expensive investigation into areas that are already conforming to specifications and procedures. However, inspection is justified if it improves safety or product performance.

Identify Managerial Responsibilities. Managerial responsibility for inspection within or outside the organization is identified next. This individual or group is responsible for planning, organizing, and reporting inspection results.

Identify Required Resources. Resources such as personnel, facilities, and equipment are assigned to any inspection. Inspectors must be trained in testing and inspection. In large facilities, inspectors conduct tests and measurements on sophisticated equipment. Inspectors also require training in inspection methods, specifications, and acceptance standards.

Inspection facilities are required for receiving, handling, and storing products. Facilities include space for shelving, storage, calibration, and test. Measurement and test equipment must also be secured.

Calibrated measurement equipment is required for performing inspection and testing. Typical measurement equipment includes nondestructive, destructive, electrical, dimensional, optical, mechanical, and pneumatic equipment.

Identify Quality Attributes. A product can have 200 or more dimensions. Some are critical, while others are major, minor, or inconsequential. The inspector must know which should be measured. Quality attributes not only include dimensional measurements, but also performance, maintainability, reliability, safety, aesthetics, usability, color, ergonomics, and design. The customer, design engineer, or manufacturing engineer identifies these on a drawing, procedure, or specification. If quality characteristics are open to interpretation, they should be explained by the design engineer.

Determine Frequency of Inspection. The frequency of inspection is based on the nature of the quality attribute inspected, risk to the producer and consumer, cost of inspection, quantities in the lot, and quality history. Let us discuss these.

A critical or major quality attribute has a greater need to be inspected than a minor quality attribute. If there is any risk to the consumer from using the product, this will justify additional inspection. If production quantities are large, the frequency of inspection will be lessened because of cost. If the product quality history is good, the **inspection level** will be reduced.

Define Inspection Methods and Acceptance Criteria. The inspector must know how quality characteristics should be measured. Once engineering indicates which characteristics need to be inspected, a procedure is written to describe how critical and major quality attributes should be measured.

The inspection method depends on the product. Even though operators may use the same piece of equipment, procedures, techniques, and knack among operators may differ. Set-up, observation, equipment handling, equipment calibration, and gaging may also differ.

The inspector must have sufficient information to determine whether a product conforms to specifications. Acceptance criteria for a product may include visual appearance, test results, and dimensional tolerances.

The inspector uses a statistical sampling plan that specifies the **acceptance number** in the sample. This number determines whether the lot will be accepted or rejected.

Report Inspection Results. Information gathered from inspection is reported to the appropriate parties. If a part is nonconforming, results are reported to the design engineer or to the supplier. A documentation trail is necessary if management wants to identify the source of a nonconformance so that corrective action can be initiated to eliminate the nonconformance. The inspection report should identify the inspector, date of inspection, work order number, item, number of lot, inspection level, and results.

INTRODUCTION TO SAMPLING

Types of Inspection

The extent of any sampling can range from none to statistical to **100% sampling.** While we briefly discuss 100% sampling, we devote the rest of this chapter to **statistical sampling.**

Highly mechanized processes sometimes use automatic test equipment to test 100% of a quality characteristic in products coming down the production line. As products move along the line, cameras or measuring equipment check critical dimensions. If a dimension is off, a machine automatically segregates nonconforming parts into a special bin. Some machines are also capable of adjusting themselves if a product dimension wanders off target.

Batches of products are 100% sampled if circumstances warrant it or if a customer requests it; 100% sampling is warranted to comply with government regulations, to satisfy special customer requirements, or to inspect one-of-a-kind products.

Comply with Government Regulations. Depending on the critical nature of a product and its intended use, the government may require 100% test, evaluation, and inspection. For example, airline luggage is 100% inspected.

Satisfy Special Customer Requirements. Special testing and 100% sampling of critical or major quality attributes may be required by a customer or regulatory authority. Critical attributes affect product safety. Major attributes affect product function.

Inspect One-of-a-Kind Products. A one-of-a-kind product is usually expensive and once in service cannot be repaired easily. For example, a satellite put into orbit cannot be replaced or repaired. To prevent premature failure, a one-of-a-kind product is inspected thoroughly during production. If a flaw is discovered, the product can be repaired immediately instead of after many operational steps or while in use.

Problems of 100% Sampling

One hundred percent sampling should be used only in special circumstances to satisfy the preceding conditions. However, there are problems with 100% sampling. Specifically it is expensive, inaccurate, impractical, and becomes a sorting procedure.

Expensive. One hundred percent sampling means that all the products of a batch or shipment are inspected, tested, and evaluated to ensure conformance to specifications. This is expensive and does not add value to the product. It requires additional inspectors, measuring equipment, trained or certified technicians, facilities, and storage space. Testing, inspection, and evaluation are especially expensive for a complex product.

Inaccurate. One hundred percent sampling is only 80% effective. Mistakes are made because inspection becomes monotonous and the inspector's attention wanders.

Impractical. Some types of testing destroy the product. Destructive tests evaluate products to failure and the products have to be scrapped. It is obviously impractical to destruct 100% of a shipment.

Sorting Procedure. One hundred percent sampling becomes a sorting procedure for an operation or vendor. Unfortunately the customer or recipient of the part is communicating that he or she does not trust the efforts of the vendor or operator.

The operator will eventually expect that his or her efforts will be checked and all nonconforming parts will be caught and sorted by the inspector. This takes responsibility for product quality off the person who should be ultimately responsible, the operator.

What Is Statistical Sampling?

Sampling or, more specifically, **acceptance sampling** is the probability-based technique of determining the **sample size** to be pulled from a population given certain requirements. The purpose of acceptance sampling is to supply information about the batch or lot from which the sample is drawn. This assumes the sample is representative of the population.

The goal of acceptance sampling is to eventually lead to some prevention technique such as statistical process control (SPC). As we have stressed throughout this text, controlling and assuring the results of the process is preferred over defect inspection. However, there are still many uses for inspection. Acceptance sampling inspection is used in sampling products in incoming, in-process, and final inspection.

In general, statistical sampling is preferred over 100% sampling because it is less expensive, more efficient, less damaging to the products, and it is a strong positive signal.

Less Expensive. Statistical sampling is less expensive than 100% inspection, for which additional space, personnel, equipment, and resources are required.

More Efficient. Statistical sampling, properly planned and implemented, should result in as good or perhaps better control of incoming material quality than 100% sampling. As discussed, 100% manual sampling is monotonous and hypnotic and results in errors.

Less Damaging to Products. A shipment can have 50 or more lots of slightly different material. For example, a shipment of small-diameter threaded pipe may contain many lots, each having a different diameter and length of pipe. Each lot is segregated, and a different sized sample is pulled from each lot depending on population size. Each time boxes are moved and ripped open, and material is inspected, the contents may be damaged.

Strong Positive Signal. Statistical sampling is an important first step toward prevention. Statistical sampling tables permit the inspector to reduce or tighten inspection. At first any inspection is loose, but over time, inspection is tightened as quality improves, until SPC or some other prevention technique can be used. This results in a strong signal to the operator or the vendor that they are responsible for the quality of their efforts and products.

Disadvantages of Sampling

Any type of inspection presents problems. In-process inspection disrupts operations because partially processed material is dumped at an inspection area until a decision is made on disposition. Inspection becomes a tollgate for a process where all material is funneled into the inspection area. Inspection areas become holding areas that are bottlenecks to the travel of material. This also disrupts the master production schedule. Also, material kept in a holding area is expensive because it is like material in inventory.

Statistical sampling is not 100% effective. There is a chance of accepting bad lots and rejecting good lots. When a small random sample is pulled from a population of products, the inspector will not check every part. Products that should be rejected will be accepted because no defective products were found in the inspected sample.

Statistical sampling entails pulling a small lot from a large population of items. The lot may not be representative of the entire population; it may not have been chosen randomly.

Statistical sampling requires extensive planning and training of personnel. An appropriate sampling plan must be chosen and an area must be set aside for inspection. Furthermore, people need to be trained in statistical sampling, testing, and measurement equipment.

INTRODUCTION TO ACCEPTANCE SAMPLING

General Sampling Procedure

Most statistical sampling plans follow a procedure where a sample of items is randomly selected from a batch, lot, or shipment. This sample is inspected for defects or defectives. If the number of defects or defectives is less than or equal to a number, the lot is accepted. This number is called the acceptance number. If the number of defectives is greater than the acceptance number, the lot is rejected.

The following example illustrates how sampling is used in quality control. A needle bearing is a critical component in a military vehicle. If the outside diameter (OD) dimension of the bearing is out of round, excessive wear generates unwanted heat and causes the transmission to freeze. The manufacturer uses statistical process techniques in monitoring and controlling the OD dimension and states that the process is in control, $C_p = 1.0$, or 99.7% conforming. However, the military wants to confirm the manufacturer's claims. The supply depot decides to sample and inspect incoming products to ensure that 99.7% or more of the product conforms to specifications.

The inspector uses a **sampling plan** and specifies **lot size,** quality level, and inspection level. The sampling plan specifies that a number of products should be pulled from the shipment for inspection. If the number of defective ball bearings is equal to or less than the minimum specified in the sampling plan, the lot is accepted. If the number of products is greater than the acceptance number, the lot is rejected.

Sampling Planning

Sampling, like inspection, should be planned. It should be ensured prior to implementation that lots are homogeneous, that samples are selected randomly, and that inspectors follow prescribed procedures and inspection is uniform.

Lots Are Homogeneous. The formation of the lots is critical if sampling is to result in meaningful information. Lots should be homogeneous in terms of being produced by the same operator, machine, raw material, and procedures. As we have stressed, causes of assignable variation should be monitored and minimized prior to inspection and sampling.

If there is substantial variation in a lot, sampling will not work. A shipment may consist of 1000 parts produced by four plants that combine their outputs into one shipment, which will have parts made by different operators using different machines and following different procedures, possibly under different conditions, with different raw materials.

It is impossible to pull a homogeneous sample under these conditions, and if nonconforming material is found, it is impossible to pinpoint its source to take corrective action in order to eliminate the root cause of the nonconformance.

Samples Are Randomly Selected. A sample should be selected randomly and be representative of the entire lot. To ensure that samples are selected randomly, each box, or bin, in a lot is assigned a number. Random numbers are obtained from a table or generated by a computer, telling the quality inspector which boxes and in what order they should be sampled and tested.

Inspectors Follow Prescribed Procedures. Inspectors should follow prescribed procedures when sampling, inspecting, or testing. It is possible after a long day to shortcut or not follow procedures. If this occurs, the risk of accepting defective products increases.

Unfortunately quality control sometimes is in an adversarial relationship with production. Production is pressured to push products out the door as quality control is inspecting and rejecting shipments. The customer may want the products immediately, or sales may have promised delivery as products are waiting to be inspected.

SAMPLE SELECTION

The selection of a sample is so important to accurate sampling that it deserves additional explanation. There are two general methods of drawing a sample from a population: probability methods and nonprobability methods. If all products in the population are represented in the sample, this is called a **probability sample.** If the selection of products in the sample relies on the subjective judgment of the inspector, the sample is called a **nonprobability sample.**

Nonprobability sampling, also called **incidental sampling,** has applications in quality. For example, before a product is mass-produced, a company may form a test group to solicit user information or assess a new product idea. This small group is not totally representative of the population in the niche market.

The main disadvantage of nonprobability sampling is that the accuracy of the results cannot be expressed and assessed mathematically. Also, items chosen in nonprobability sampling do not have an equal chance of being selected. However, nonprobability sampling offers several advantages. It is less expensive, less time consuming, and more convenient than probability sampling.

The rest of this chapter focuses on probability, or statistical, sampling. In **statistical sampling** the inspector or analyst tries to select a random sample with the same characteristics as the entire population.

There are several ways to select a random, uniform, and representative sample from a heterogeneous population and make a reliable quality decision: random, systematic, or stratified sampling.

Random Sampling. In **random sampling,** products are selected randomly from a population. A random sample is one where each item in the population has an equal chance of being selected. The theory of sampling is based on the concept that the sample is representative of the population.

As much as possible, lots should consist of products of a single type, grade, class, size, and composition. Furthermore, lots should be manufactured under essentially the same conditions and at the same time. The reason for this uniformity is to minimize any variation so that a reliable accept/reject quality decision can be made.

One way to select a random sample is to use a random number table or a computer that generates random numbers that refer to items in the population. Before this is done, all of the items in the population are assigned a number. Sometimes a random sample is pulled from a phone book using a similar method, where each person in the book has a preassigned number or the computer simply selects a person's name. Each name in the phone book has an equal chance of being selected.

Systematic Sampling. Random sampling is difficult and expensive if the population of items is large. **Systematic sampling** is a logical method to simplify mass random sampling. A large population of items can be organized in a systematic manner so that samples are random and representative of the population.

For example, if a machine continuously stamps 1000 products per hour, the operator does not have time to refer to a computer to obtain a random, representative sample of all 1000 products. The operator's attention is focused on monitoring and adjusting his or her machine. The operator knows that stamped products go into a bin and that each bin holds 50 homogeneous parts. It is easier for the operator to randomly pull products from each fifth bin than to worry about pulling products from each hour's production.

Stratified Sampling. **Stratified sampling** is another method of simplifying random selection of products. In stratified sampling, the total population is divided into subgroups, or strata. Each subpopulation in a subgroup is sampled independently of the other subpopulations.

The purpose behind stratification is to select a representative sample that will reflect the characteristics of the population of products in the lot, or shipment. The shipment may consist of 2-inch bolts threaded by different machines, operated by different operators using different methods. Also, the metal may not be consistent. Probably the bolts from each threading machine are packaged separately from those produced by the other machines. So by stratifying, the inspector hopes the products in the sample represent those in the shipment.

OPERATING CHARACTERISTIC CURVES

In any statistical sampling, there is always the chance that a ''good'' lot will be rejected and a ''bad'' lot will be accepted. **Operating characteristic (OC) curves** indicate different levels of risk associated with performing a certain type of inspection. Specifically, OC curves indicate the probability of accepting a lot for various percent or fraction defective rates.

Typical OC Curve

By interpreting an OC curve, a quality analyst obtains the probability of accepting a lot with a given percent defective value $100p$. You may recall that 8% defective is equivalent to a 0.08 fraction defective. The OC curve in Figure 7.1 represents a typical sampling plan. It indicates that there is a 50% probability of accepting a 20% defective lot. Also, there is a 25% probability of accepting a 40% defective lot.

Several elements of the OC curve should first be explained.

☐ 100% conforming material is accepted 100% of the time. The OC curve indicates that the probability of accepting 100% conforming material is 1.00, or 100%.

☐ 100% defective material is rejected 100% of the time. The OC curve indicates that the probability of rejecting 100% defective material is 1.00, or 100%.

Ideal OC Curve

The shape of an ideal OC curve differs from that of the typical OC curve (Figure 7.2). The optimum OC curve accepts all low percent-defective lots and rejects all

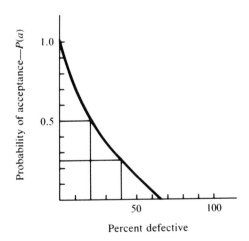

FIGURE 7.1
Typical OC curve.

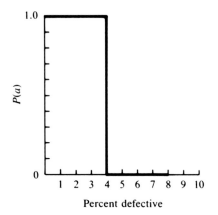

FIGURE 7.2
Ideal OC curve.

lots that have high percent defectives. According to our ideal curve, a shipment is rejected every time its percent defective is greater than or to the right of a given value. The optimum OC curve illustrates this point. Let us assume the quality analyst wants to accept all lots less than 4% defective and reject all lots greater than 4% defective. All lots less than 4% defective have a 100% (1.0) probability of acceptance, and all lots greater than 4% have a 0% probability of acceptance. Thus if the percent defective of the product is less than the given value, the product is accepted all the time. In real life this is seldom realized. This type of curve can be closely approximated by a typical OC curve.

Constructing an OC Curve

Four factors, percent defective, population (lot) size, sample size, and acceptance number, define the shape of the OC curve. In the sampling tables (see Tables 7.2–7.10 at the end of this chapter) these are designated as follows:

$$p = \text{percent defective}$$
$$N = \text{population size}$$
$$n = \text{sample size}$$
$$c = \text{acceptance number}$$

By changing these parameters, OC curves can be developed for an infinite number of sampling situations. First the probability of accepting a lot at several levels of percent defective is determined statistically. Depending on the lot size, sample size, and percent defective, an OC curve is then constructed by substituting the appropriate values into the hypergeometric, binomial, or Poisson distribution equations and calculating and plotting the probabilities on the OC curve. This process is lengthy and time consuming. It is much easier to use predetermined tables to construct the curves.

Fortunately the Poisson probability distribution is the easiest to use in constructing the OC curves. However, before using the Poisson distribution, the quality analyst must determine that the lot size is 10 times larger than the sample size and the percent defective is small, 2% or less. Then a probability table is set up. Probabilities are either calculated or derived using the Poisson table given in Appendix B. The latter is easier, as shown in the following.

The cumulative probability for the four acceptance values $c = 0, 1, 2,$ and 3 is calculated using the following Poisson equation:

$$P(x) = \frac{(np)^x e^{-np}}{x!}$$

This is the same Poisson equation discussed in Chapter 4. The above equation is different, however, because np replaces λ. Previously, the symbol λ represented the average rate at which events occur. Similarly, np is the conventional notation referring to the average value of the expected number of defectives.

For $N = 1000$, $n = 100$, and $p = 0.02$,

$$P(0) = \frac{(2)^0 e^{-2}}{0!} = 0.135$$

$$P(1) = \frac{(2)^1 e^{-2}}{1!} = 0.271$$

$$P(2) = \frac{(2)^2 e^{-2}}{2!} = 0.271$$

$$P(3) = \frac{(2)^3 e^{-2}}{3!} = 0.180$$

Then

$$P(\text{total}) = P(0) + P(1) + P(2) + P(3)$$
$$= 0.135 + 0.271 + 0.271 + 0.180 = 0.857$$

Calculating probabilities using the Poisson distribution is tedious. It can be facilitated by using the Poisson distribution values given in Appendix B. This table lists the probabilities of accepting c or fewer defects for various values of percent defective np.

The horizontal line gives the percent defective values np, and the vertical line lists the acceptance values c. For example, given $N = 1000$, $n = 100$, $c = 0$, and $p = 0.02$,

$$np = 100 \times 0.02 = 2.0$$

Knowing that $np = 2$ and $c = 0$, we refer to the Poisson distribution values in Appendix B and find 0.135, or 13.5%.

Cumulative values for various acceptance numbers $c = 0, 1, 2, \ldots$ are shown in parentheses. The following calculation can be verified by referring to

Appendix B, where for $np = 2$ and $c = 3$ we find a cumulative probability of 0.857.

EXAMPLE   Given the sampling condition shown below, construct a probability table. Sampling plan: $N = 1000$, $n = 100$, $c = 2$. Using Appendix B the following is found. See Figure 7.3.

Solution

p	np	$P(a)$
0.01	1	0.920
0.02	2	0.677
0.03	3	0.423
0.04	4	0.238
0.05	5	0.125
0.08	8	0.014

EXAMPLE   Find the probability of accepting a shipment for which $p = 0.01$, $n = 90$, $N = 2000$, and $c = 3$.

Solution
We know $np = 0.01 \times 90 = 0.9$ and $c = 3$. By referring to the Poisson distribution values in Appendix B, the solution is 0.987, or 98.7%.

Producer's and Consumer's Risk

Sampling inspection has the intrinsic risk that good lots may be rejected or bad lots may be accepted. In any sampling plan, these risks must be considered.

☐ **Producer's (alpha) risk** is the probability that a "good" lot will be rejected by the sampling plan. This means the lot contains an acceptable percent defective.

☐ **Consumer's (beta) risk** is the probability that a "bad" lot will be accepted by the sampling plan. This means the lot has an unacceptable percent defective.*

Sampling plans are sometimes described in terms of producer's risk and consumer's risk. In some sampling plans, producer's risk is established at 5.0 percent defective and consumer's risk is set at 10.0 percent defective. However, these risk values will change to reflect the needs of the organization.

The relationship between the two types of risks is illustrated in Figure 7.3. The curve shows that for an acceptable quality level (AQL) of 1%, there is an 8% risk of rejecting the entire lot. Also from the consumer's perspective, there is a 10% risk of accepting the lot that is 5.2% defective.

---

* J. M. Juran, *Quality Control Handbook*, 3rd ed. (McGraw-Hill, New York, 1979), pp. 24-8 to 24-16.

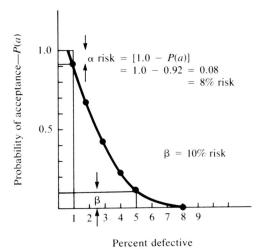

Sampling plan

$n = 100$
$c = 2$

p	np	$p(a)$
0.01	1	0.920
0.02	2	0.677
0.03	3	0.423
0.04	4	0.238
0.05	5	0.125
0.08	8	0.014

α risk $= [1.0 - P(a)]$
$= 1.0 - 0.92 = 0.08$
$= 8\%$ risk

$\beta = 10\%$ risk

Probability of acceptance—$P(a)$

Percent defective

FIGURE 7.3
Producer's and consumer's risks.

MIL-STD 105D

Military standard 105D* is one of the most extensively used sampling plans. This plan, called inspection by attributes, consists of three plans: single, double, and multiple sampling.

The sampling plans work similarly. Given certain criteria, a sampling plan tells the inspector how many products to pull from the population. The inspector then counts the presence or absence of some quality attribute in each product of a sample selected from a shipment. A product is classified as defective or nondefective based on the results of the inspection.

If a product has different types of defects, the number and type of defects in each product are counted, and a determination to accept or reject the product is

* MIL-STD 105D, "Sampling Procedures and Tables for Inspection by Attributes," Superintendent of Documents, Washington, DC, 1963.

made. A product may have three minor defects and still be accepted, while a product with one critical defect would be rejected. Finally, an accept or reject decision is made of the entire lot or shipment based on the number of rejects in the sample. Depending on the sampling plan, there may be one or more samples.

In single, double, or multiple sampling, important pieces of data have to be specified. These are the same data that were used to construct the OC curves. The first is the number of products N in the batch from which the sample is to be drawn. The second is the number of products n in the random sample. The third is the acceptance number c, which is the number of allowable defective products in the sample. If the batch has more than a number c of defective products, the batch is rejected.

Single Sampling. In **single sampling** the inspector pulls a sample of products based on the size of the batch or lot and the quality level required. The inspector accepts or rejects the batch based on criteria found in sampling tables. The quality level is the amount of protection the inspector wants. A higher level of protection means that for the same number of products in the batch, the inspector pulls a larger sized sample.

Double Sampling. As the name implies, the inspector pulls two separate samples in **double sampling.** In the first sample the inspector pulls fewer products than in single sampling to maintain the same level of protection. Based on the results of the inspection, the inspector either accepts the lot or pulls a second sample, which determines whether the lot is accepted or rejected. By using a smaller first sample, double sampling permits the lot to be accepted or rejected based on fewer units than in single sampling.

Multiple Sampling. **Multiple sampling** allows for another sample to be drawn if the first two samples have not been sufficient to accept the batch. In multiple sampling a maximum number of samples is drawn until an accept or reject decision is finally reached.

Sampling Plan Procedure

Now we demonstrate how these different plans are used. MIL-STD 105D indicates the number of products from each shipment, batch, or lot to be inspected. The acceptance and rejection numbers are listed in the master tables (see Tables 7.2–7.10 at the end of this chapter).

To determine the number of products to select, the quality inspector must know or specify the following: inspection level, lot size, AQL, and sampling condition.

Inspection Level. Three general inspection levels, I, II, and III, are displayed in Table 7.1. They provide different levels of assurance. Level III involves the most inspection, which results in a higher level of assurance, more discrimination, and higher inspection costs.

The inspector or quality analyst changes the inspection level based on the type of product and the cost of inspection. A complex or critical product may require level III inspection, while an inexpensive commodity would justify level I inspection. When inspection starts, it is usually at a Level II.

Lot Size. Lot size is the number of products in a lot, batch, or shipment. A letter designates the lot size for a particular level of inspection (Table 7.1). Depending on the type of sampling plan, the inspector next goes to the master tables (Tables 7.2–7.10).

AQL. AQL is the maximum percent defective, or the maximum number of defects per 100 units, that can be considered satisfactory as a process average. It is expressed either as percent defective or as defects per 100 products. MIL-STD 105D sampling plans use AQL to determine the quality levels. AQL values are listed in the top rows of the master tables (Tables 7.2–7.10).

To determine the number of products to select, the inspector matches the inspection letter, given in Table 7.1, with the AQL number from the master tables (Tables 7.2–7.10). If there is no acceptance number, the master table indicates through vertical arrows to go to another inspection letter.

When a customer specifies an AQL for a shipment of incoming products, the AQL indicates that the acceptance plan will accept the majority of the supplier's shipments if the process average of defects per 100 units is no greater than the AQL value.

The customer or supplier may specify AQLs for individual quality characteristics or for products. A complex product may have 10 or more quality characteristics being inspected. The characteristics may be classified as critical, major, or minor. Depending on the criticality of the quality characteristic, the AQLs may be different. A critical defect in an important quality characteristic will result in hazardous or unsafe conditions. It has a higher AQL than a minor defect.

TABLE 7.1 (Sample size code letters)

Lot Size	General Inspection Levels		
	I	II	III
2–8	A	A	B
9–15	A	B	C
16–25	B	C	D
26–50	C	D	E
51–90	C	E	F
91–150	D	F	G
151–280	E	G	H
281–500	F	H	J
501–1200	G	J	K
1201–3200	H	K	L
3201–10000	J	L	M
10001–35000	K	M	N

Sampling plans are arranged so that the probability of acceptance at the specified AQL depends on the sample size. For a given AQL, the probability of acceptance is generally higher for a larger sized sample than for a smaller sized sample.

Sampling Condition. The degree of inspection, normal, reduced, or tightened, determines the relationship between the lot size and the sample size. **Tightened inspection** specifies a larger sample size than reduced or normal inspection. Usually if a product has no quality history, normal inspection, is used. Normal inspection continues unchanged for each class of defect or percent defective until sufficient quality history has been obtained.

As the inspector obtains a history of product quality, he or she changes the degree of inspection. If two of five lots are rejected, the inspector goes to tightened inspection. On the other hand, if five consecutive lots are accepted on normal inspection, **reduced inspection** is instituted.

Sampling Plans

The following examples show how to use the MIL-STD 105D master tables (Tables 7.2–7.10 at the end of the chapter). In the examples, inspection level (II), lot size (1000), and AQL (2.5%) stay the same, while the sampling plans and the sampling conditions change.

The examples illustrate three different sample plans: single, double, and multiple sampling. Within each example, three sampling conditions are illustrated, normal, tightened, and reduced.

The mechanics for using the master tables are fairly straightforward. The quality inspector or analyst first finds the sample size code letter from Table 7.1. Since the lot size (1000) is constant throughout the examples, the sample size code letter J, obtained from Table 7.1, is the same in all examples.

Within the master tables there are vertical arrows, advising the inspector to go to the next higher or lower sample size code letter, which means that the sample size changes accordingly. If an arrow points downward, a larger sample size is required to make a statistically reliable sampling decision. If an arrow points upward, a smaller sample size can be used to make a similar reliable decision.

SINGLE SAMPLING PLAN EXAMPLES

Given:

Sampling plan	Single
Inspection level	II
Lot size	1000
AQL	2.5%
Sampling condition	Normal

Solution

From Table 7.2 the plan specifies that 80 pieces are pulled and inspected. The AQL (2.5%) is constant throughout all examples. At the intersection of the AQL column of 2.5% and the row of code letter J, the inspector finds the accept (Ac) number to be 5 and the reject (Re) number to be 6.

This means that out of a lot or batch of 1000 products, a random sample of 80 pieces is inspected. If 5 or less defective products are found, the lot is accepted. If 6 or more defective products are found, the lot is rejected.

Given:

Sampling plan	Single
Inspection level	II
Lot size	1000
AQL	2.5%
Sampling condition	Tightened

Solution

The sample size code letter J indicates again that the sample size should be 80 pieces. In Table 7.3 the inspector finds the accept number (Ac) to be 3 and the reject number (Re) to be 4. If 3 or less defective products are found, the lot is accepted, and if 4 or more defective products are found, the lot is rejected.

Given:

Sampling plan	Single
Inspection level	II
Lot size	1000
AQL	2.5%
Sampling condition	Reduced

Solution

Across the row for sample size code letter J the sample size is now 32. In Table 7.4, under AQL 2.5%, the accept number (Ac) is 2 and the reject number (Re) is 5. If 2 or fewer defective pieces are pulled from the lot, the lot is accepted. If 5 or more defective pieces are pulled, the lot is rejected. If 3 or 4 defectives are found, the lot is accepted upon the condition that the normal sampling condition resumes in the next inspection.

DOUBLE SAMPLING PLAN EXAMPLES

Given:

Sampling plan	Double
Inspection level	II
Lot size	1000
AQL	2.5%
Sampling condition	Normal

Solution

In Table 7.5 the following accept and reject numbers are given for the above conditions:

| | | Cumulative | | |
Sample	Sample Size	Sample Size	Ac	Re
First	50	50	2	5
Second	50	100	6	7

From a lot of 1000 products, a random sample of 50 is first pulled and inspected. If 2 or less are defective products, the lot is accepted. If 5 or more are defective products, the lot is rejected. If the lot has 3 or 4 defective products, a second sample of 50 is pulled and inspected. If the total number of defective products from both samples (the cumulative sample size is 100) is 6 or less, the lot is accepted. If the total number is 7 or more, the lot is rejected.

Given:

Sampling plan	Double
Inspection level	II
Lot size	1000
AQL	2.5%
Sampling condition	Tightened

Solution
In Table 7.6 the following accept and reject numbers are given for the above conditions:

| | | Cumulative | | |
Sample	Sample Size	Sample Size	Ac	Re
First	50	50	1	4
Second	50	100	4	5

Again from a lot of 1000 products, a random sample of 50 is pulled and inspected. If 1 or less are defective products, the lot is accepted. If 4 or more are defective products, the lot is rejected. If the lot has 2 or 3 defective products, a second sample of 50 is pulled and inspected. If the total number of defective products from both samples is 4 or less, the lot is accepted. If the total number is 5 or more, the lot is rejected.

Given:

Sampling plan	Double
Inspection level	II
Lot size	1000
AQL	2.5%
Sampling condition	Reduced

Solution
In Table 7.7 the following accept and reject numbers are given for the above conditions:

	Sample Size	Cumulative Sample Size	Ac	Re
Sample				
First	20	20	0	4
Second	20	40	3	6

In the first sample, 20 products are now pulled and inspected. If 0 products are defective, the lot is accepted, and if 4 or more are defective, the lot is rejected. If the number of defectives is 1, 2, or 3, a second sample is inspected. If the total number of defectives in both samples is 3 or less, the lot is accepted. If the total number of defectives is 6 or more, the lot is rejected and normal inspection is reinstated. If the total number of defectives is 4 or 5, the lot is accepted and normal inspection resumes.

MULTIPLE SAMPLING PLAN EXAMPLES

Given:

Sampling plan	Multiple
Inspection level	II
Lot size	1000
AQL	2.5%
Sampling condition	Normal

In the multiple sampling tables, two new symbols are introduced, # and ††. The symbol # means that the lot cannot be accepted based on the results of sampling only one lot, even if there are no defectives, because the sample size is too small. However, if the sample has the required number or more defectives, the lot can be rejected. A number between the accept and reject numbers means that an additional sample has to be pulled and inspected. The symbol †† means that either the corresponding double sampling plan or the multiple sampling plan should be used.

Multiple sampling plans are supplied for normal, tightened, and reduced inspection in Tables 7.8–7.10. The difference between theses tables and those for double sampling is that an inspector can pull an initial smaller sample and make a quick accept/reject decision. This works well if a lot is known to be consistently acceptable or unacceptable. However, if the lot is known to be marginal, the inspector may have to pull additional samples until an accept or reject decision can be made.

Multiple sampling is similar to double sampling. In this case, if the first sample of 20 products has 4 or more defectives, the lot is rejected. No more samples have to be pulled. The # symbol indicates, however, that the lot cannot be accepted, even if there are no defectives, until a second lot is pulled and inspected.

The procedure then becomes similar to double sampling. Samples are pulled and inspected until finally an accept or reject decision is made. If the number of defectives is between the accept and reject numbers, another sample is pulled and inspected. The accept and reject numbers represent the total defectives from the cumulative samples.

Solution

Sample	Sample Size	Cumulative Sample Size	Ac	Re
First	20	20	#	4
Second	20	40	1	5
Third	20	60	2	6
Fourth	20	80	3	7
Fifth	20	100	5	8
Sixth	20	120	7	9
Seventh	20	140	9	10

Given:

Sampling plan	Multiple
Inspection level	II
Lot size	1000
AQL	2.5%
Sampling condition	Tightened

Solution
No explanation is required, because sampling and analysis are similar to normal multiple sampling.

Sample	Sample Size	Cumulative Sample Size	Ac	Re
First	20	20	#	3
Second	20	40	0	3
Third	20	60	1	4
Fourth	20	80	2	5
Fifth	20	100	3	6
Sixth	20	120	4	6
Seventh	20	140	6	7

Given:

Sampling plan	Multiple
Inspection level	II
Lot size	1000
AQL	2.5%
Sampling condition	Reduced

Solution
Again, sampling and analysis are similar to the above example.

Sample	Sample Size	Cumulative Sample Size	Ac	Re
First	8	8	#	3
Second	8	16	0	4
Third	8	24	0	5
Fourth	8	32	1	6
Fifth	8	40	2	7
Sixth	8	48	3	7
Seventh	8	56	4	8

Evaluation of Sampling Plans

Sampling plans were developed to respond to different inspection requirements. The choice of which plan to use is based on material quality and sampling costs. If the product quality is marginal for a required level of protection, double or multiple sampling plans should be used. In these sampling plans the inspector continues to inspect a shipment until an accept or reject decision is made. If the lot is marginal, the inspector cannot make a quick determination without pulling a larger representative sample. However, if the supplier has a history of producing known acceptable or rejectable products, a smaller sample will result in a quick accept or reject decision.

Also, the average sample size is less for multiple plans than for double and single plans. If the quality of material is known to be acceptable, the cost per unit for single sampling is less than that for double or multiple sampling. If material quality is questionable, double or multiple sampling plans will require the inspector to pull more samples, which increases the cost of sampling.

Continuous Improvement

The goal of any acceptance sampling is to improve process and product quality. A plan may have been developed 5 or 10 years ago to meet a desired quality level. Several events may have occurred. The levels of risk and control may have changed over time. The original product may have been modified so that previous quality levels no longer apply. Inspection levels should change to reflect changing needs or desired quality improvement.

Sampling inspection should be a dynamic process of continuous improvement until prevention can eventually evolve. Inspection is initially used to obtain a level of quality history. Once this is obtained, the operator or supplier is notified of the results and asked to improve the process.

TABLE 7.2 Single sampling plans for normal inspection

Acceptable Quality Levels (normal inspection)

(Each cell gives the acceptance/rejection pair as "Ac Re". ↓ = use first sampling plan below arrow; ↑ = use first sampling plan above arrow.)

Code	Sample size	0.010	0.015	0.025	0.040	0.065	0.10	0.15	0.25	0.40	0.65	1.0	1.5	2.5	4.0	6.5	10	15	25	40	65	100	150	250	400	650	1000
A	2	↓	↓	↓	↓	↓	↓	↓	↓	↓	↓	↓	↓	↓	↓	↓	↓	0 1	1 2	2 3	3 4	5 6	7 8	10 11	14 15	21 22	30 31
B	3	↓	↓	↓	↓	↓	↓	↓	↓	↓	↓	↓	↓	↓	↓	↓	0 1	1 2	2 3	3 4	5 6	7 8	10 11	14 15	21 22	30 31	44 45
C	5	↓	↓	↓	↓	↓	↓	↓	↓	↓	↓	↓	↓	↓	↓	0 1	1 2	2 3	3 4	5 6	7 8	10 11	14 15	21 22	30 31	44 45	↑
D	8	↓	↓	↓	↓	↓	↓	↓	↓	↓	↓	↓	↓	↓	0 1	1 2	2 3	3 4	5 6	7 8	10 11	14 15	21 22	30 31	44 45	↑	↑
E	13	↓	↓	↓	↓	↓	↓	↓	↓	↓	↓	↓	↓	0 1	1 2	2 3	3 4	5 6	7 8	10 11	14 15	21 22	30 31	44 45	↑	↑	↑
F	20	↓	↓	↓	↓	↓	↓	↓	↓	↓	↓	↓	0 1	1 2	2 3	3 4	5 6	7 8	10 11	14 15	21 22	30 31	44 45	↑	↑	↑	↑
G	32	↓	↓	↓	↓	↓	↓	↓	↓	↓	↓	0 1	1 2	2 3	3 4	5 6	7 8	10 11	14 15	21 22	30 31	44 45	↑	↑	↑	↑	↑
H	50	↓	↓	↓	↓	↓	↓	↓	↓	↓	0 1	1 2	2 3	3 4	5 6	7 8	10 11	14 15	21 22	30 31	44 45	↑	↑	↑	↑	↑	↑
J	80	↓	↓	↓	↓	↓	↓	↓	↓	0 1	1 2	2 3	3 4	5 6	7 8	10 11	14 15	21 22	30 31	44 45	↑	↑	↑	↑	↑	↑	↑
K	125	↓	↓	↓	↓	↓	↓	↓	0 1	1 2	2 3	3 4	5 6	7 8	10 11	14 15	21 22	30 31	44 45	↑	↑	↑	↑	↑	↑	↑	↑
L	200	↓	↓	↓	↓	↓	↓	0 1	1 2	2 3	3 4	5 6	7 8	10 11	14 15	21 22	30 31	44 45	↑	↑	↑	↑	↑	↑	↑	↑	↑
M	315	↓	↓	↓	↓	↓	0 1	1 2	2 3	3 4	5 6	7 8	10 11	14 15	21 22	30 31	44 45	↑	↑	↑	↑	↑	↑	↑	↑	↑	↑
N	500	↓	↓	↓	↓	0 1	1 2	2 3	3 4	5 6	7 8	10 11	14 15	21 22	30 31	44 45	↑	↑	↑	↑	↑	↑	↑	↑	↑	↑	↑
P	800	↓	↓	↓	0 1	1 2	2 3	3 4	5 6	7 8	10 11	14 15	21 22	30 31	44 45	↑	↑	↑	↑	↑	↑	↑	↑	↑	↑	↑	↑
Q	1250	↓	↓	0 1	1 2	2 3	3 4	5 6	7 8	10 11	14 15	21 22	30 31	44 45	↑	↑	↑	↑	↑	↑	↑	↑	↑	↑	↑	↑	↑
R	2000	↓	0 1	1 2	2 3	3 4	5 6	7 8	10 11	14 15	21 22	30 31	44 45	↑	↑	↑	↑	↑	↑	↑	↑	↑	↑	↑	↑	↑	↑

⇩ = Use first sampling plan below arrow. If sample size equals, or exceeds, lot or batch size, do 100 percent inspection.

⇧ = Use first sampling plan above arrow.

Ac = Acceptance number.

Re = Rejection number.

* Reproduced from MIL-STD 105D, "Sampling Procedures and Tables for Inspection by Attributes," Superintendent of Documents, Washington, DC, 1963.

TABLE 7.3 Single sampling plans for tightened inspection

Acceptable Quality Levels (tightened inspection). Each cell shows **Ac Re** (acceptance number, rejection number). ↓ = Use first sampling plan below arrow. ↑ = Use first sampling plan above arrow.

Sample size code letter	Sample size	0.010	0.015	0.025	0.040	0.065	0.10	0.15	0.25	0.40	0.65	1.0	1.5	2.5	4.0	6.5	10	15	25	40	65	100	150	250	400	650	1000
A	2	↓	↓	↓	↓	↓	↓	↓	↓	↓	↓	↓	↓	↓	↓	↓	↓	↓	0 1	1 2	2 3	3 4	5 6	8 9	12 13	18 19	27 28
B	3	↓	↓	↓	↓	↓	↓	↓	↓	↓	↓	↓	↓	↓	↓	↓	↓	0 1	1 2	2 3	3 4	5 6	8 9	12 13	18 19	27 28	41 42
C	5	↓	↓	↓	↓	↓	↓	↓	↓	↓	↓	↓	↓	↓	↓	↓	0 1	1 2	2 3	3 4	5 6	8 9	12 13	18 19	27 28	41 42	↑
D	8	↓	↓	↓	↓	↓	↓	↓	↓	↓	↓	↓	↓	↓	↓	0 1	1 2	2 3	3 4	5 6	8 9	12 13	18 19	27 28	41 42	↑	↑
E	13	↓	↓	↓	↓	↓	↓	↓	↓	↓	↓	↓	↓	↓	0 1	1 2	2 3	3 4	5 6	8 9	12 13	18 19	27 28	41 42	↑	↑	↑
F	20	↓	↓	↓	↓	↓	↓	↓	↓	↓	↓	↓	↓	0 1	1 2	2 3	3 4	5 6	8 9	12 13	18 19	27 28	41 42	↑	↑	↑	↑
G	32	↓	↓	↓	↓	↓	↓	↓	↓	↓	↓	↓	0 1	1 2	2 3	3 4	5 6	8 9	12 13	18 19	27 28	41 42	↑	↑	↑	↑	↑
H	50	↓	↓	↓	↓	↓	↓	↓	↓	↓	↓	0 1	1 2	2 3	3 4	5 6	8 9	12 13	18 19	27 28	41 42	↑	↑	↑	↑	↑	↑
J	80	↓	↓	↓	↓	↓	↓	↓	↓	↓	0 1	1 2	2 3	3 4	5 6	8 9	12 13	18 19	27 28	41 42	↑	↑	↑	↑	↑	↑	↑
K	125	↓	↓	↓	↓	↓	↓	↓	↓	0 1	1 2	2 3	3 4	5 6	8 9	12 13	18 19	27 28	41 42	↑	↑	↑	↑	↑	↑	↑	↑
L	200	↓	↓	↓	↓	↓	↓	↓	0 1	1 2	2 3	3 4	5 6	8 9	12 13	18 19	27 28	41 42	↑	↑	↑	↑	↑	↑	↑	↑	↑
M	315	↓	↓	↓	↓	↓	↓	0 1	1 2	2 3	3 4	5 6	8 9	12 13	18 19	27 28	41 42	↑	↑	↑	↑	↑	↑	↑	↑	↑	↑
N	500	↓	↓	↓	↓	↓	0 1	1 2	2 3	3 4	5 6	8 9	12 13	18 19	27 28	41 42	↑	↑	↑	↑	↑	↑	↑	↑	↑	↑	↑
P	800	↓	↓	↓	↓	0 1	1 2	2 3	3 4	5 6	8 9	12 13	18 19	27 28	41 42	↑	↑	↑	↑	↑	↑	↑	↑	↑	↑	↑	↑
Q	1250	↓	↓	↓	0 1	1 2	2 3	3 4	5 6	8 9	12 13	18 19	27 28	41 42	↑	↑	↑	↑	↑	↑	↑	↑	↑	↑	↑	↑	↑
R	2000	↓	↓	0 1	1 2	2 3	3 4	5 6	8 9	12 13	18 19	27 28	41 42	↑	↑	↑	↑	↑	↑	↑	↑	↑	↑	↑	↑	↑	↑
S	3150	↓	0 1	1 2	2 3	3 4	5 6	8 9	12 13	18 19	27 28	41 42	↑	↑	↑	↑	↑	↑	↑	↑	↑	↑	↑	↑	↑	↑	↑

↓ = Use first sampling plan below arrow. If sample size equals or exceeds lot or batch size, do 100 percent inspection.

↑ = Use first sampling plan above arrow.

Ac = Acceptance number.

Re = Rejection number.

* Reproduced from MIL-STD 105D, "Sampling Procedures and Tables for Inspection by Attributes," Superintendent of Documents, Washington, DC, 1963.

TABLE 7.4 Single sampling plans for reduced inspection

Sample size code letter	Sample size	Acceptable Quality Levels (reduced inspection)†																									
		0.010	0.015	0.025	0.040	0.065	0.10	0.15	0.25	0.40	0.65	1.0	1.5	2.5	4.0	6.5	10	15	25	40	65	100	150	250	400	650	1000
		Ac Re	Ac Re	Ac Re	Ac Re	Ac Re	Ac Re	Ac Re	Ac Re	Ac Re	Ac Re	Ac Re	Ac Re	Ac Re	Ac Re	Ac Re	Ac Re	Ac Re	Ac Re	Ac Re	Ac Re	Ac Re	Ac Re	Ac Re	Ac Re	Ac Re	Ac Re
A	2	↓	↓	↓	↓	↓	↓	↓	↓	↓	↓	↓	↓	↓	↓	0 1	↓	↓	1 2	2 3	3 4	5 6	7 8	10 11	14 15	21 22	30 31
B	2	↓	↓	↓	↓	↓	↓	↓	↓	↓	↓	↓	↓	↓	0 1	↑	↓	0 2	1 3	2 4	3 5	5 6	7 8	10 11	14 15	21 22	30 31
C	2	↓	↓	↓	↓	↓	↓	↓	↓	↓	↓	↓	↓	0 1	↑	↓	0 2	1 3	1 4	2 5	3 6	5 8	7 10	10 13	14 17	21 24	↑
D	3	↓	↓	↓	↓	↓	↓	↓	↓	↓	↓	↓	0 1	↑	↓	0 2	1 3	1 4	2 5	3 6	5 8	7 10	10 13	14 17	21 24	↑	↑
E	5	↓	↓	↓	↓	↓	↓	↓	↓	↓	↓	0 1	↑	↓	0 2	1 3	1 4	2 5	3 6	5 8	7 10	10 13	14 17	21 24	↑	↑	↑
F	8	↓	↓	↓	↓	↓	↓	↓	↓	↓	0 1	↑	↓	0 2	1 3	1 4	2 5	3 6	5 8	7 10	10 13	↑	↑	↑	↑	↑	↑
G	13	↓	↓	↓	↓	↓	↓	↓	↓	0 1	↑	↓	0 2	1 3	1 4	2 5	3 6	5 8	7 10	10 13	↑	↑	↑	↑	↑	↑	↑
H	20	↓	↓	↓	↓	↓	↓	↓	0 1	↑	↓	0 2	1 3	1 4	2 5	3 6	5 8	7 10	10 13	↑	↑	↑	↑	↑	↑	↑	↑
J	32	↓	↓	↓	↓	↓	↓	0 1	↑	↓	0 2	1 3	1 4	2 5	3 6	5 8	7 10	10 13	↑	↑	↑	↑	↑	↑	↑	↑	↑
K	50	↓	↓	↓	↓	↓	0 1	↑	↓	0 2	1 3	1 4	2 5	3 6	5 8	7 10	10 13	↑	↑	↑	↑	↑	↑	↑	↑	↑	↑
L	80	↓	↓	↓	↓	0 1	↑	↓	0 2	1 3	1 4	2 5	3 6	5 8	7 10	10 13	↑	↑	↑	↑	↑	↑	↑	↑	↑	↑	↑
M	125	↓	↓	↓	0 1	↑	↓	0 2	1 3	1 4	2 5	3 6	5 8	7 10	10 13	↑	↑	↑	↑	↑	↑	↑	↑	↑	↑	↑	↑
N	200	↓	↓	0 1	↑	↓	0 2	1 3	1 4	2 5	3 6	5 8	7 10	10 13	↑	↑	↑	↑	↑	↑	↑	↑	↑	↑	↑	↑	↑
P	315	↓	0 1	↑	↓	0 2	1 3	1 4	2 5	3 6	5 8	7 10	10 13	↑	↑	↑	↑	↑	↑	↑	↑	↑	↑	↑	↑	↑	↑
Q	500	0 1	↑	↓	0 2	1 3	1 4	2 5	3 6	5 8	7 10	10 13	↑	↑	↑	↑	↑	↑	↑	↑	↑	↑	↑	↑	↑	↑	↑
R	800	↑	↑	0 2	1 3	1 4	2 5	3 6	5 8	7 10	10 13	↑	↑	↑	↑	↑	↑	↑	↑	↑	↑	↑	↑	↑	↑	↑	↑

↓ = Use first sampling plan below arrow. If sample size equals or exceeds lot or batch size, do 100 percent inspection.
↑ = Use first sampling plan above arrow.
Ac = Acceptance number.
Re = Rejection number.
† = If the acceptance number has been exceeded, but the rejection number has not been reached, accept the lot, but reinstate normal inspection (see 10.1.4).

* Reproduced from MIL-STD 105D, "Sampling Procedures and Tables for Inspection by Attributes," Superintendent of Documents, Washington, DC, 1963.

TABLE 7.5 Double sampling plans for normal inspection

Acceptable Quality Levels (normal inspection)

The data cells below give, for each sample size code letter, the **First** sample Acceptance/Rejection numbers (Ac Re) and the cumulative **Second** sample Acceptance/Rejection numbers (Ac Re) at each AQL. In the arrow regions: ↓ = "Use first sampling plan below arrow. If sample size equals or exceeds lot or batch size do 100 percent inspection"; ↑ = "Use first sampling plan above arrow"; • = "Use corresponding single sampling plan (or alternatively, use double sampling plan below, where available)." Ac = Acceptance number; Re = Rejection number.

Code	Sample	Sample size	Cum. sample size	0.010	0.015	0.025	0.040	0.065	0.10	0.15	0.25	0.40	0.65	1.0	1.5	2.5	4.0	6.5	10	15	25	40	65	100	150	250	400	650	1000
A				↓	↓	↓	↓	↓	↓	↓	↓	↓	↓	↓	↓	↓	↓	↓	↓	↓	•	•	•	•	•	•	•	•	•
B	First	2	2	↓	↓	↓	↓	↓	↓	↓	↓	↓	↓	↓	↓	↓	↓	↓	•	0 2	0 3	1 4	2 5	3 7	5 9	7 11	11 16	17 22	25 31
	Second	2	4																	1 2	3 4	4 5	6 7	8 9	12 13	18 19	26 27	37 38	56 57
C	First	3	3	↓	↓	↓	↓	↓	↓	↓	↓	↓	↓	↓	↓	↓	↓	•	0 2	0 3	1 4	2 5	3 7	5 9	7 11	11 16	17 22	25 31	↑
	Second	3	6																1 2	3 4	4 5	6 7	8 9	12 13	18 19	26 27	37 38	56 57	
D	First	5	5	↓	↓	↓	↓	↓	↓	↓	↓	↓	↓	↓	↓	↓	•	0 2	0 3	1 4	2 5	3 7	5 9	7 11	11 16	17 22	25 31	↑	↑
	Second	5	10															1 2	3 4	4 5	6 7	8 9	12 13	18 19	26 27	37 38	56 57		
E	First	8	8	↓	↓	↓	↓	↓	↓	↓	↓	↓	↓	↓	↓	•	0 2	0 3	1 4	2 5	3 7	5 9	7 11	11 16	17 22	25 31	↑	↑	↑
	Second	8	16														1 2	3 4	4 5	6 7	8 9	12 13	18 19	26 27	37 38	56 57			
F	First	13	13	↓	↓	↓	↓	↓	↓	↓	↓	↓	↓	↓	•	0 2	0 3	1 4	2 5	3 7	5 9	7 11	11 16	17 22	25 31	↑	↑	↑	↑
	Second	13	26													1 2	3 4	4 5	6 7	8 9	12 13	18 19	26 27	37 38	56 57				
G	First	20	20	↓	↓	↓	↓	↓	↓	↓	↓	↓	↓	•	0 2	0 3	1 4	2 5	3 7	5 9	7 11	11 16	17 22	25 31	↑	↑	↑	↑	↑
	Second	20	40												1 2	3 4	4 5	6 7	8 9	12 13	18 19	26 27	37 38	56 57					
H	First	32	32	↓	↓	↓	↓	↓	↓	↓	↓	↓	•	0 2	0 3	1 4	2 5	3 7	5 9	7 11	11 16	17 22	25 31	↑	↑	↑	↑	↑	↑
	Second	32	64											1 2	3 4	4 5	6 7	8 9	12 13	18 19	26 27	37 38	56 57						
J	First	50	50	↓	↓	↓	↓	↓	↓	↓	↓	•	0 2	0 3	1 4	2 5	3 7	5 9	7 11	11 16	17 22	25 31	↑	↑	↑	↑	↑	↑	↑
	Second	50	100										1 2	3 4	4 5	6 7	8 9	12 13	18 19	26 27	37 38	56 57							
K	First	80	80	↓	↓	↓	↓	↓	↓	↓	•	0 2	0 3	1 4	2 5	3 7	5 9	7 11	11 16	17 22	25 31	↑	↑	↑	↑	↑	↑	↑	↑
	Second	80	160									1 2	3 4	4 5	6 7	8 9	12 13	18 19	26 27	37 38	56 57								
L	First	125	125	↓	↓	↓	↓	↓	↓	•	0 2	0 3	1 4	2 5	3 7	5 9	7 11	11 16	17 22	25 31	↑	↑	↑	↑	↑	↑	↑	↑	↑
	Second	125	250								1 2	3 4	4 5	6 7	8 9	12 13	18 19	26 27	37 38	56 57									
M	First	200	200	↓	↓	↓	↓	↓	•	0 2	0 3	1 4	2 5	3 7	5 9	7 11	11 16	17 22	25 31	↑	↑	↑	↑	↑	↑	↑	↑	↑	↑
	Second	200	400							1 2	3 4	4 5	6 7	8 9	12 13	18 19	26 27	37 38	56 57										
N	First	315	315	↓	↓	↓	↓	•	0 2	0 3	1 4	2 5	3 7	5 9	7 11	11 16	17 22	25 31	↑	↑	↑	↑	↑	↑	↑	↑	↑	↑	↑
	Second	315	630						1 2	3 4	4 5	6 7	8 9	12 13	18 19	26 27	37 38	56 57											
P	First	500	500	↓	↓	↓	•	0 2	0 3	1 4	2 5	3 7	5 9	7 11	11 16	17 22	25 31	↑	↑	↑	↑	↑	↑	↑	↑	↑	↑	↑	↑
	Second	500	1000					1 2	3 4	4 5	6 7	8 9	12 13	18 19	26 27	37 38	56 57												
Q	First	800	800	↓	↓	•	0 2	0 3	1 4	2 5	3 7	5 9	7 11	11 16	17 22	25 31	↑	↑	↑	↑	↑	↑	↑	↑	↑	↑	↑	↑	↑
	Second	800	1600				1 2	3 4	4 5	6 7	8 9	12 13	18 19	26 27	37 38	56 57													
R	First	1250	1250	↓	•	0 2	0 3	1 4	2 5	3 7	5 9	7 11	11 16	17 22	25 31	↑	↑	↑	↑	↑	↑	↑	↑	↑	↑	↑	↑	↑	↑
	Second	1250	2500			1 2	3 4	4 5	6 7	8 9	12 13	18 19	26 27	37 38	56 57														

↓ = Use first sampling plan below arrow. If sample size equals or exceeds lot or batch size do 100 percent inspection
↑ = Use first sampling plan above arrow
Ac = Acceptance number
Re = Rejection number
• = Use corresponding single sample plan (or alternatively, use double sampling plan below, where available)

* Reproduced from MIL-STD 105D, "Sampling Procedures and Tables for Inspection by Attributes," Superintendent of Documents, Washington, DC, 1963.

TABLE 7.6 Double sampling plans for tightened inspection

Acceptable Quality Levels (tightened inspection)

Sample size code letter	Sample	Sample size	Cumulative sample size	0.010 Ac	0.010 Re	0.015 Ac	0.015 Re	0.025 Ac	0.025 Re	0.040 Ac	0.040 Re	0.065 Ac	0.065 Re	0.10 Ac	0.10 Re	0.15 Ac	0.15 Re	0.25 Ac	0.25 Re	0.40 Ac	0.40 Re	0.65 Ac	0.65 Re	1.0 Ac	1.0 Re	1.5 Ac	1.5 Re	2.5 Ac	2.5 Re	4.0 Ac	4.0 Re	6.5 Ac	6.5 Re	10 Ac	10 Re	15 Ac	15 Re	25 Ac	25 Re	40 Ac	40 Re	65 Ac	65 Re	100 Ac	100 Re	150 Ac	150 Re	250 Ac	250 Re	400 Ac	400 Re	650 Ac	650 Re	1000 Ac	1000 Re
A				↓		↓		↓		↓		↓		↓		↓		↓		↓		↓		↓		↓		↓		↓		↓		↓		↓		•		•		•		•		•		•		•		•	
B	First	2	2	↓		↓		↓		↓		↓		↓		↓		↓		↓		↓		↓		↓		↓		↓		↓		↓		•		0	2	0	3	1	4	2	5	3	7	6	10	9	14	15	20	23	24
B	Second	2	4																																			1	2	3	4	4	5	6	7	11	12	15	16	23	24	34	35	52	53
C	First	3	3	↓		↓		↓		↓		↓		↓		↓		↓		↓		↓		↓		↓		↓		↓		•		0	2	0	3	1	4	2	5	3	7	6	10	9	14	15	20	23	24	↑	
C	Second	3	6																															1	2	3	4	4	5	6	7	11	12	15	16	23	24	34	35	52	53		
D	First	5	5	↓		↓		↓		↓		↓		↓		↓		↓		↓		↓		↓		↓		↓		•		0	2	0	3	1	4	2	5	3	7	6	10	9	14	15	20	23	24	↑		↑	
D	Second	5	10																													1	2	3	4	4	5	6	7	11	12	15	16	23	24	34	35	52	53				
E	First	8	8	↓		↓		↓		↓		↓		↓		↓		↓		↓		↓		↓		↓		•		0	2	0	3	1	4	2	5	3	7	6	10	9	14	15	20	23	24	↑		↑		↑	
E	Second	8	16																											1	2	3	4	4	5	6	7	11	12	15	16	23	24	34	35	52	53						
F	First	13	13	↓		↓		↓		↓		↓		↓		↓		↓		↓		↓		↓		•		0	2	0	3	1	4	2	5	3	7	6	10	9	14	15	20	23	24	↑		↑		↑		↑	
F	Second	13	26																									1	2	3	4	4	5	6	7	11	12	15	16	23	24	34	35	52	53								
G	First	20	20	↓		↓		↓		↓		↓		↓		↓		↓		↓		↓		•		0	2	0	3	1	4	2	5	3	7	6	10	9	14	15	20	23	24	↑		↑		↑		↑		↑	
G	Second	20	40																					1	2	3	4	4	5	6	7	11	12	15	16	23	24	34	35	52	53										
H	First	32	32	↓		↓		↓		↓		↓		↓		↓		↓		↓		•		0	2	0	3	1	4	2	5	3	7	6	10	9	14	15	20	23	24	↑		↑		↑		↑		↑		↑	
H	Second	32	64																	1	2	3	4	4	5	6	7	11	12	15	16	23	24	34	35	52	53												
J	First	50	50	↓		↓		↓		↓		↓		↓		↓		↓		•		0	2	0	3	1	4	2	5	3	7	6	10	9	14	15	20	23	24	↑		↑		↑		↑		↑		↑		↑	
J	Second	50	100															1	2	3	4	4	5	6	7	11	12	15	16	23	24	34	35	52	53														
K	First	80	80	↓		↓		↓		↓		↓		↓		↓		•		0	2	0	3	1	4	2	5	3	7	6	10	9	14	15	20	23	24	↑		↑		↑		↑		↑		↑		↑		↑	
K	Second	80	160													1	2	3	4	4	5	6	7	11	12	15	16	23	24	34	35	52	53																
L	First	125	125	↓		↓		↓		↓		↓		↓		•		0	2	0	3	1	4	2	5	3	7	6	10	9	14	15	20	23	24	↑		↑		↑		↑		↑		↑		↑		↑		↑	
L	Second	125	250											1	2	3	4	4	5	6	7	11	12	15	16	23	24	34	35	52	53																		
M	First	200	200	↓		↓		↓		↓		↓		•		0	2	0	3	1	4	2	5	3	7	6	10	9	14	15	20	23	24	↑		↑		↑		↑		↑		↑		↑		↑		↑		↑	
M	Second	200	400									1	2	3	4	4	5	6	7	11	12	15	16	23	24	34	35	52	53																				
N	First	315	315	↓		↓		↓		↓		•		0	2	0	3	1	4	2	5	3	7	6	10	9	14	15	20	23	24	↑		↑		↑		↑		↑		↑		↑		↑		↑		↑		↑	
N	Second	315	630							1	2	3	4	4	5	6	7	11	12	15	16	23	24	34	35	52	53																						
P	First	500	500	↓		↓		↓		•		0	2	0	3	1	4	2	5	3	7	6	10	9	14	15	20	23	24	↑		↑		↑		↑		↑		↑		↑		↑		↑		↑		↑		↑	
P	Second	500	1000					1	2	3	4	4	5	6	7	11	12	15	16	23	24	34	35	52	53																								
Q	First	800	800	↓		↓		•		0	2	0	3	1	4	2	5	3	7	6	10	9	14	15	20	23	24	↑		↑		↑		↑		↑		↑		↑		↑		↑		↑		↑		↑		↑	
Q	Second	800	1600			1	2	3	4	4	5	6	7	11	12	15	16	23	24	34	35	52	53																										
R	First	1250	1250	↓		•		0	2	0	3	1	4	2	5	3	7	6	10	9	14	15	20	23	24	↑		↑		↑		↑		↑		↑		↑		↑		↑		↑		↑		↑		↑		↑	
R	Second	1250	2500	1	2	3	4	4	5	6	7	11	12	15	16	23	24	34	35	52	53																												
S	First	2000	2000	•		0	2	0	3	1	4	2	5	3	7	6	10	9	14	15	20	23	24	↑		↑		↑		↑		↑		↑		↑		↑		↑		↑		↑		↑		↑		↑		↑	
S	Second	2000	4000	1	2	3	4	4	5	6	7	11	12	15	16	23	24	34	35	52	53																												

⇩ Use first sampling plan below arrow. If sample size equals or exceeds lot or batch size, do 100 percent inspection.

⇧ Use first sampling plan above arrow.

Ac Acceptance number

Re Rejection number

• Use corresponding single sampling plan (or alternatively, use double sampling plan below, where available)

* Reproduced from MIL-STD 105D, "Sampling Procedures and Tables for Inspection by Attributes," Superintendent of Documents, Washington, DC, 1963.

TABLE 7.7 Double sampling plans for reduced inspection

Acceptable Quality Levels (reduced inspection)†

Ac = Acceptance number Re = Rejection number

Note on the data grid: the body of this chart is a diagonal (staircase) arrangement. In each cell the upper pair of numbers is the First-sample "Ac Re" and the lower pair is the Second-sample "Ac Re". Arrow symbols and the "*" symbol occupy the cells outside the numeric band, as explained in the legend. The reading below gives, for each code letter, the First / Second "Ac Re" values in the numbered cells, with ↓ / ↑ / * marking the non-numeric cells.

Sample size code letter	Sample	Sample size	Cumulative sample size	Numbered cells (AQL : First Ac Re / Second Ac Re)
A				* (single-sampling plan used; arrows otherwise)
B				* (single-sampling plan used; arrows otherwise)
C				* (single-sampling plan used; arrows otherwise)
D	First	2	2	4.0: * ; 6.5: 0 2 / 0 2 ; 10: 0 3 / 1 4 ; 15: 0 4 / 1 5 ; 25: 1 4 / 4 5 ; 40: 1 5 / 4 6 ; 65: 2 6 / 5 7 ; 100: 3 8 / 8 12 ; 150: 5 12 / 10 16 ; 250: 7 18 / 11 26 ; 400: 11 22 / 17 30 ; 650–1000: ↑ ; ≤2.5: ↓
	Second	2	4	
E	First	3	3	2.5: * ; 4.0: 0 2 / 0 2 ; 6.5: 0 3 / 1 4 ; 10: 0 4 / 1 5 ; 15: 1 4 / 4 5 ; 25: 1 5 / 4 6 ; 40: 2 6 / 5 7 ; 65: 3 8 / 8 12 ; 100: 5 12 / 10 16 ; 150: 7 18 / 11 26 ; 250: 11 22 / 17 30 ; 400–1000: ↑ ; ≤1.5: ↓
	Second	3	6	
F	First	5	5	1.5: * ; 2.5: 0 2 / 0 2 ; 4.0: 0 3 / 1 4 ; 6.5: 0 4 / 1 5 ; 10: 1 4 / 4 5 ; 15: 1 5 / 4 6 ; 25: 2 6 / 5 7 ; 40: 3 8 / 8 12 ; 65: 5 12 / 10 16 ; 100: 7 18 / 11 26 ; 150: 11 22 / 17 30 ; 250–1000: ↑ ; ≤1.0: ↓
	Second	5	10	
G	First	8	8	1.0: * ; 1.5: 0 2 / 0 2 ; 2.5: 0 3 / 1 4 ; 4.0: 0 4 / 1 5 ; 6.5: 1 4 / 4 5 ; 10: 1 5 / 4 6 ; 15: 2 6 / 5 7 ; 25: 3 8 / 8 12 ; 40: 5 12 / 10 16 ; 65: 7 18 / 11 26 ; 100: 11 22 / 17 30 ; 150–1000: ↑ ; ≤0.65: ↓
	Second	8	16	
H	First	13	13	0.65: * ; 1.0: 0 2 / 0 2 ; 1.5: 0 3 / 1 4 ; 2.5: 0 4 / 1 5 ; 4.0: 1 4 / 4 5 ; 6.5: 1 5 / 4 6 ; 10: 2 6 / 5 7 ; 15: 3 8 / 8 12 ; 25: 5 12 / 10 16 ; 40: 7 18 / 11 26 ; 65: 11 22 / 17 30 ; 100–1000: ↑ ; ≤0.40: ↓
	Second	13	26	
J	First	20	20	0.40: * ; 0.65: 0 2 / 0 2 ; 1.0: 0 3 / 1 4 ; 1.5: 0 4 / 1 5 ; 2.5: 1 4 / 4 5 ; 4.0: 1 5 / 4 6 ; 6.5: 2 6 / 5 7 ; 10: 3 8 / 8 12 ; 15: 5 12 / 10 16 ; 25: 7 18 / 11 26 ; 40: 11 22 / 17 30 ; 65–1000: ↑ ; ≤0.25: ↓
	Second	20	40	
K	First	32	32	0.25: * ; 0.40: 0 2 / 0 2 ; 0.65: 0 3 / 1 4 ; 1.0: 0 4 / 1 5 ; 1.5: 1 4 / 4 5 ; 2.5: 1 5 / 4 6 ; 4.0: 2 6 / 5 7 ; 6.5: 3 8 / 8 12 ; 10: 5 12 / 10 16 ; 15: 7 18 / 11 26 ; 25: 11 22 / 17 30 ; 40–1000: ↑ ; ≤0.15: ↓
	Second	32	64	
L	First	50	50	0.15: * ; 0.25: 0 2 / 0 2 ; 0.40: 0 3 / 1 4 ; 0.65: 0 4 / 1 5 ; 1.0: 1 4 / 4 5 ; 1.5: 1 5 / 4 6 ; 2.5: 2 6 / 5 7 ; 4.0: 3 8 / 8 12 ; 6.5: 5 12 / 10 16 ; 10: 7 18 / 11 26 ; 15: 11 22 / 17 30 ; 25–1000: ↑ ; ≤0.10: ↓
	Second	50	100	
M	First	80	80	0.10: * ; 0.15: 0 2 / 0 2 ; 0.25: 0 3 / 1 4 ; 0.40: 0 4 / 1 5 ; 0.65: 1 4 / 4 5 ; 1.0: 1 5 / 4 6 ; 1.5: 2 6 / 5 7 ; 2.5: 3 8 / 8 12 ; 4.0: 5 12 / 10 16 ; 6.5: 7 18 / 11 26 ; 10: 11 22 / 17 30 ; 15–1000: ↑ ; ≤0.065: ↓
	Second	80	160	
N	First	125	125	0.065: * ; 0.10: 0 2 / 0 2 ; 0.15: 0 3 / 1 4 ; 0.25: 0 4 / 1 5 ; 0.40: 1 4 / 4 5 ; 0.65: 1 5 / 4 6 ; 1.0: 2 6 / 5 7 ; 1.5: 3 8 / 8 12 ; 2.5: 5 12 / 10 16 ; 4.0: 7 18 / 11 26 ; 6.5: 11 22 / 17 30 ; 10–1000: ↑ ; ≤0.040: ↓
	Second	125	250	
P	First	200	200	0.040: * ; 0.065: 0 2 / 0 2 ; 0.10: 0 3 / 1 4 ; 0.15: 0 4 / 1 5 ; 0.25: 1 4 / 4 5 ; 0.40: 1 5 / 4 6 ; 0.65: 2 6 / 5 7 ; 1.0: 3 8 / 8 12 ; 1.5: 5 12 / 10 16 ; 2.5: 7 18 / 11 26 ; 4.0: 11 22 / 17 30 ; 6.5–1000: ↑ ; ≤0.025: ↓
	Second	200	400	
Q	First	315	315	0.025: * ; 0.040: 0 2 / 0 2 ; 0.065: 0 3 / 1 4 ; 0.10: 0 4 / 1 5 ; 0.15: 1 4 / 4 5 ; 0.25: 1 5 / 4 6 ; 0.40: 2 6 / 5 7 ; 0.65: 3 8 / 8 12 ; 1.0: 5 12 / 10 16 ; 1.5: 7 18 / 11 26 ; 2.5: 11 22 / 17 30 ; 4.0–1000: ↑ ; ≤0.015: ↓
	Second	315	630	
R	First	500	500	0.015: * ; 0.025: 0 2 / 0 2 ; 0.040: 0 3 / 1 4 ; 0.065: 0 4 / 1 5 ; 0.10: 1 4 / 4 5 ; 0.15: 1 5 / 4 6 ; 0.25: 2 6 / 5 7 ; 0.40: 3 8 / 8 12 ; 0.65: 5 12 / 10 16 ; 1.0: 7 18 / 11 26 ; 1.5: 11 22 / 17 30 ; 2.5–1000: ↑ ; 0.010: ↓
	Second	500	1000	

↓ = Use first sampling plan below arrow. If sample size equals or exceeds lot or batch size, do 100 percent inspection

↑ = Use first sampling plan above arrow

Ac = Acceptance number
Re = Rejection number

* = Use corresponding single sampling plan below (or alternatively use double sampling plan below, when available)

† = If, after the second sample, the acceptance number has been exceeded, but the rejection number has not been reached, accept the lot, but reinstate normal inspection (see 10.14)

Reproduced from MIL–STD 105D, "Sampling Procedures and Tables for Inspection by Attributes," Superintendent of Documents, Washington, DC, 1963.

185

TABLE 7.8 Multiple sampling plans for normal inspection

Acceptable Quality Levels (normal inspection). Entries give the Acceptance number (Ac) and Rejection number (Re) for each successive sample. Sample size code letters A, B and C have no multiple sampling plan (arrows only). For AQL columns to the left of the tabulated values use the first sampling plan below the arrow (larger code letter); for AQL columns to the right use the first sampling plan above the arrow, or the corresponding single / double sampling plan (‡ + / ‡).

Sample size code letter D — Sample size 2

Sample	Sample size	Cumulative sample size	25 (Ac Re)	40 (Ac Re)	65 (Ac Re)	100 (Ac Re)	150 (Ac Re)	250 (Ac Re)	400 (Ac Re)
First	2	2	* 4	0 4	0 5	1 7	2 9	4 12	6 16
Second	2	4	1 5	1 6	3 8	4 10	7 14	11 19	17 27
Third	2	6	2 6	3 8	6 10	8 13	13 19	19 27	29 39
Fourth	2	8	3 7	5 10	8 13	12 17	19 25	27 34	40 49
Fifth	2	10	5 8	7 11	11 15	17 20	25 29	36 40	53 58
Sixth	2	12	7 9	10 12	14 17	21 23	31 33	45 47	66 68
Seventh	2	14	9 10	13 14	18 19	25 26	37 38	53 54	77 78

Sample size code letter E — Sample size 3

Sample	Sample size	Cumulative sample size	6.5 (Ac Re)	10 (Ac Re)	15 (Ac Re)	25 (Ac Re)	40 (Ac Re)	65 (Ac Re)	100 (Ac Re)	150 (Ac Re)	250 (Ac Re)
First	3	3	* 2	* 3	* 4	0 4	0 5	1 7	2 9	4 12	6 16
Second	3	6	0 3	0 3	1 5	1 6	3 8	4 10	7 14	11 19	17 27
Third	3	9	0 3	1 4	2 6	3 8	6 10	8 13	13 19	19 27	29 39
Fourth	3	12	1 4	2 5	3 7	5 10	8 13	12 17	19 25	27 34	40 49
Fifth	3	15	2 5	3 6	5 8	7 11	11 15	17 20	25 29	36 40	53 58
Sixth	3	18	3 5	4 6	7 9	10 12	14 17	21 23	31 33	45 47	66 68
Seventh	3	21	4 5	6 7	9 10	13 14	18 19	25 26	37 38	53 54	77 78

Sample size code letter F — Sample size 5

Sample	Sample size	Cumulative sample size	4.0 (Ac Re)	6.5 (Ac Re)	10 (Ac Re)	15 (Ac Re)	25 (Ac Re)	40 (Ac Re)	65 (Ac Re)	100 (Ac Re)	150 (Ac Re)
First	5	5	* 2	* 3	* 4	0 4	0 5	1 7	2 9	4 12	6 16
Second	5	10	0 3	0 3	1 5	1 6	3 8	4 10	7 14	11 19	17 27
Third	5	15	0 3	1 4	2 6	3 8	6 10	8 13	13 19	19 27	29 39
Fourth	5	20	1 4	2 5	3 7	5 10	8 13	12 17	19 25	27 34	40 49
Fifth	5	25	2 5	3 6	5 8	7 11	11 15	17 20	25 29	36 40	53 58
Sixth	5	30	3 5	4 6	7 9	10 12	14 17	21 23	31 33	45 47	66 68
Seventh	5	35	4 5	6 7	9 10	13 14	18 19	25 26	37 38	53 54	77 78

Sample size code letter G — Sample size 8

Sample	Sample size	Cumulative sample size	2.5 (Ac Re)	4.0 (Ac Re)	6.5 (Ac Re)	10 (Ac Re)	15 (Ac Re)	25 (Ac Re)	40 (Ac Re)	65 (Ac Re)	100 (Ac Re)
First	8	8	* 2	* 3	* 4	0 4	0 5	1 7	2 9	4 12	6 16
Second	8	16	0 3	0 3	1 5	1 6	3 8	4 10	7 14	11 19	17 27
Third	8	24	0 3	1 4	2 6	3 8	6 10	8 13	13 19	19 27	29 39
Fourth	8	32	1 4	2 5	3 7	5 10	8 13	12 17	19 25	27 34	40 49
Fifth	8	40	2 5	3 6	5 8	7 11	11 15	17 20	25 29	36 40	53 58
Sixth	8	48	3 5	4 6	7 9	10 12	14 17	21 23	31 33	45 47	66 68
Seventh	8	56	4 5	6 7	9 10	13 14	18 19	25 26	37 38	53 54	77 78

Sample size code letter H — Sample size 13

Sample	Sample size	Cumulative sample size	1.5 (Ac Re)	2.5 (Ac Re)	4.0 (Ac Re)	6.5 (Ac Re)	10 (Ac Re)	15 (Ac Re)	25 (Ac Re)	40 (Ac Re)	65 (Ac Re)
First	13	13	* 2	* 3	* 4	0 4	0 5	1 7	2 9	4 12	6 16
Second	13	26	0 3	0 3	1 5	1 6	3 8	4 10	7 14	11 19	17 27
Third	13	39	0 3	1 4	2 6	3 8	6 10	8 13	13 19	19 27	29 39
Fourth	13	52	1 4	2 5	3 7	5 10	8 13	12 17	19 25	27 34	40 49
Fifth	13	65	2 5	3 6	5 8	7 11	11 15	17 20	25 29	36 40	53 58
Sixth	13	78	3 5	4 6	7 9	10 12	14 17	21 23	31 33	45 47	66 68
Seventh	13	91	4 5	6 7	9 10	13 14	18 19	25 26	37 38	53 54	77 78

Sample size code letter J — Sample size 20

Sample	Sample size	Cumulative sample size	1.0 (Ac Re)	1.5 (Ac Re)	2.5 (Ac Re)	4.0 (Ac Re)	6.5 (Ac Re)	10 (Ac Re)	15 (Ac Re)	25 (Ac Re)	40 (Ac Re)
First	20	20	* 2	* 3	* 4	0 4	0 5	1 7	2 9	4 12	6 16
Second	20	40	0 3	0 3	1 5	1 6	3 8	4 10	7 14	11 19	17 27
Third	20	60	0 3	1 4	2 6	3 8	6 10	8 13	13 19	19 27	29 39
Fourth	20	80	1 4	2 5	3 7	5 10	8 13	12 17	19 25	27 34	40 49
Fifth	20	100	2 5	3 6	5 8	7 11	11 15	17 20	25 29	36 40	53 58
Sixth	20	120	3 5	4 6	7 9	10 12	14 17	21 23	31 33	45 47	66 68
Seventh	20	140	4 5	6 7	9 10	13 14	18 19	25 26	37 38	53 54	77 78

Sample size code letters A, B and C have no multiple sampling plan tabulated (use first sampling plan below the arrow).

Legend:

↓ Use first sampling plan below arrow (refer to continuation of table on following page, when necessary). If sample size equals or exceeds lot or batch size, do 100 percent inspection.
↑ Use first sampling plan above arrow.
Ac Acceptance number.
Re Rejection number.
‡ + Use corresponding single sampling plan (or alternatively, use multiple sampling plan below, where available).
‡ Use corresponding double sampling plan (or alternatively, use multiple sampling plan below, where available).
* Acceptance not permitted at this sample size.

TABLE 7.8 Continued

Acceptable Quality Levels (normal inspection)

Cells below show the cumulative acceptance/rejection numbers as **Ac Re**.
Legend: **↓** = use first sampling plan below arrow; **↑** = use first sampling plan above arrow; ***** = use corresponding single sampling plan; **‡** = acceptance not permitted at this sample size.

AQL columns to the left of those shown (0.010, 0.015, 0.025) are "↓" (use first sampling plan below), and AQL columns to the right of those shown (15, 25, 40, 65, 100, 150, 250, 400, 650, 1000) are "↑" (use first sampling plan above). For code letter R the "*" falls at AQL 0.025.

Code letter	Sample	Sample size	Cumulative sample size	0.040	0.065	0.10	0.15	0.25	0.40	0.65	1.0	1.5	2.5	4.0	6.5	10
K	First	32	32	↓	↓	↓	↓	↓	*	‡ 2	‡ 3	‡ 4	0 4	0 5	1 7	2 9
	Second	32	64							0 2	0 3	1 5	1 6	3 8	4 10	7 14
	Third	32	96							0 2	0 4	2 6	3 8	6 10	8 13	13 19
	Fourth	32	128							0 3	1 5	3 7	5 10	8 13	12 17	19 25
	Fifth	32	160							1 3	2 6	5 8	7 11	11 15	17 20	25 29
	Sixth	32	192							1 3	3 6	7 9	10 12	14 17	21 23	31 33
	Seventh	32	224							2 3	4 7	9 10	13 14	18 19	25 26	37 38
L	First	50	50	↓	↓	↓	↓	*	‡ 2	‡ 3	‡ 4	0 4	0 5	1 7	2 9	↑
	Second	50	100						0 2	0 3	1 5	1 6	3 8	4 10	7 14	
	Third	50	150						0 2	0 4	2 6	3 8	6 10	8 13	13 19	
	Fourth	50	200						0 3	1 5	3 7	5 10	8 13	12 17	19 25	
	Fifth	50	250						1 3	2 6	5 8	7 11	11 15	17 20	25 29	
	Sixth	50	300						1 3	3 6	7 9	10 12	14 17	21 23	31 33	
	Seventh	50	350						2 3	4 7	9 10	13 14	18 19	25 26	37 38	
M	First	80	80	↓	↓	↓	*	‡ 2	‡ 3	‡ 4	0 4	0 5	1 7	2 9	↑	↑
	Second	80	160					0 2	0 3	1 5	1 6	3 8	4 10	7 14		
	Third	80	240					0 2	0 4	2 6	3 8	6 10	8 13	13 19		
	Fourth	80	320					0 3	1 5	3 7	5 10	8 13	12 17	19 25		
	Fifth	80	400					1 3	2 6	5 8	7 11	11 15	17 20	25 29		
	Sixth	80	480					1 3	3 6	7 9	10 12	14 17	21 23	31 33		
	Seventh	80	560					2 3	4 7	9 10	13 14	18 19	25 26	37 38		
N	First	125	125	↓	↓	*	‡ 2	‡ 3	‡ 4	0 4	0 5	1 7	2 9	↑	↑	↑
	Second	125	250				0 2	0 3	1 5	1 6	3 8	4 10	7 14			
	Third	125	375				0 2	0 4	2 6	3 8	6 10	8 13	13 19			
	Fourth	125	500				0 3	1 5	3 7	5 10	8 13	12 17	19 25			
	Fifth	125	625				1 3	2 6	5 8	7 11	11 15	17 20	25 29			
	Sixth	125	750				1 3	3 6	7 9	10 12	14 17	21 23	31 33			
	Seventh	125	875				2 3	4 7	9 10	13 14	18 19	25 26	37 38			
P	First	200	200	↓	*	‡ 2	‡ 3	‡ 4	0 4	0 5	1 7	2 9	↑	↑	↑	↑
	Second	200	400			0 2	0 3	1 5	1 6	3 8	4 10	7 14				
	Third	200	600			0 2	0 4	2 6	3 8	6 10	8 13	13 19				
	Fourth	200	800			0 3	1 5	3 7	5 10	8 13	12 17	19 25				
	Fifth	200	1000			1 3	2 6	5 8	7 11	11 15	17 20	25 29				
	Sixth	200	1200			1 3	3 6	7 9	10 12	14 17	21 23	31 33				
	Seventh	200	1400			2 3	4 7	9 10	13 14	18 19	25 26	37 38				
Q	First	315	315	*	‡ 2	‡ 3	‡ 4	0 4	0 5	1 7	2 9	↑	↑	↑	↑	↑
	Second	315	630		0 2	0 3	1 5	1 6	3 8	4 10	7 14					
	Third	315	945		0 2	0 4	2 6	3 8	6 10	8 13	13 19					
	Fourth	315	1260		0 3	1 5	3 7	5 10	8 13	12 17	19 25					
	Fifth	315	1575		1 3	2 6	5 8	7 11	11 15	17 20	25 29					
	Sixth	315	1890		1 3	3 6	7 9	10 12	14 17	21 23	31 33					
	Seventh	315	2205		2 3	4 7	9 10	13 14	18 19	25 26	37 38					
R	First	500	500	‡ 2	‡ 3	‡ 4	0 4	0 5	1 7	2 9	↑	↑	↑	↑	↑	↑
	Second	500	1000	0 2	0 3	1 5	1 6	3 8	4 10	7 14						
	Third	500	1500	0 2	0 4	2 6	3 8	6 10	8 13	13 19						
	Fourth	500	2000	0 3	1 5	3 7	5 10	8 13	12 17	19 25						
	Fifth	500	2500	1 3	2 6	5 8	7 11	11 15	17 20	25 29						
	Sixth	500	3000	1 3	3 6	7 9	10 12	14 17	21 23	31 33						
	Seventh	500	3500	2 3	4 7	9 10	13 14	18 19	25 26	37 38						

↓ Use first sampling plan below arrow. If sample size equals or exceeds lot or batch size, do 100 percent inspection.
↑ Use first sampling plan above arrow (refer to preceding page, when necessary).
Ac = Acceptance number
Re = Rejection number
* = Use corresponding single sampling plan (or alternatively, use multiple sampling plan below, where available).
‡ = Acceptance not permitted at this sample size

* Reproduced from MIL-STD 105D, ''Sampling Procedures and Tables for Inspection by Attributes,'' Superintendent of Documents, Washington, DC, 1963.

TABLE 7.9 Multiple sampling plans for tightened inspection

Acceptable Quality Levels (tightened inspection)

Sample size code letter	Sample	Sample size	Cumulative sample size	AQL data (Ac / Re across 0.010 … 1000)
A				(use first sampling plan below arrow)
B				(use first sampling plan below arrow)
C				(use first sampling plan below arrow)
D	First	2	2	
	Second	2	4	
	Third	2	6	
	Fourth	2	8	
	Fifth	2	10	
	Sixth	2	12	
	Seventh	2	14	
E	First	3	3	
	Second	3	6	
	Third	3	9	
	Fourth	3	12	
	Fifth	3	15	
	Sixth	3	18	
	Seventh	3	21	
F	First	5	5	
	Second	5	10	
	Third	5	15	
	Fourth	5	20	
	Fifth	5	25	
	Sixth	5	30	
	Seventh	5	35	
G	First	8	8	
	Second	8	16	
	Third	8	24	
	Fourth	8	32	
	Fifth	8	40	
	Sixth	8	48	
	Seventh	8	56	
H	First	13	13	
	Second	13	26	
	Third	13	39	
	Fourth	13	52	
	Fifth	13	65	
	Sixth	13	78	
	Seventh	13	91	
J	First	20	20	
	Second	20	40	
	Third	20	60	
	Fourth	20	80	
	Fifth	20	100	
	Sixth	20	120	
	Seventh	20	140	

AQL column headings (each with Ac and Re sub-columns): 0.010, 0.015, 0.025, 0.040, 0.065, 0.10, 0.15, 0.25, 0.40, 0.65, 1.0, 1.5, 2.5, 4.0, 6.5, 10, 15, 25, 40, 65, 100, 150, 250, 400, 650, 1000.

Representative Ac/Re block read at the high-AQL columns (e.g. code D, AQL 400):
6 15, 16 25, 26 36, 37 46, 49 55, 61 64, 72 73.

Legend:

◇⇩ = Use first sampling plan below arrow (refer to continuation of table on following page, when necessary). If sample size equals or exceeds lot or batch size, do 100 percent inspection.
⇧ = Use first sampling plan above arrow.
Ac = Acceptance number.
Re = Rejection number.
* = Use corresponding single sampling plan (or alternatively, use multiple sampling plan below, where available).
++ = Use corresponding double sampling plan (or alternatively, use multiple sampling plan below, where available).
# = Acceptance not permitted at this sample size

188

TABLE 7.9 Continued

Acceptable Quality Levels (tightened inspection)

Sample size code letter	Sample	Sample size	Cumulative sample size
K	First	32	32
	Second	32	64
	Third	32	96
	Fourth	32	128
	Fifth	32	160
	Sixth	32	192
	Seventh	32	224
L	First	50	50
	Second	50	100
	Third	50	150
	Fourth	50	200
	Fifth	50	250
	Sixth	50	300
	Seventh	50	350
M	First	80	80
	Second	80	160
	Third	80	240
	Fourth	80	320
	Fifth	80	400
	Sixth	80	480
	Seventh	80	560
N	First	125	125
	Second	125	250
	Third	125	375
	Fourth	125	500
	Fifth	125	625
	Sixth	125	750
	Seventh	125	875
P	First	200	200
	Second	200	400
	Third	200	600
	Fourth	200	800
	Fifth	200	1000
	Sixth	200	1200
	Seventh	200	1400
Q	First	315	315
	Second	315	630
	Third	315	945
	Fourth	315	1260
	Fifth	315	1575
	Sixth	315	1890
	Seventh	315	2205
R	First	500	500
	Second	500	1000
	Third	500	1500
	Fourth	500	2000
	Fifth	500	2500
	Sixth	500	3000
	Seventh	500	3500
S	First	800	800
	Second	800	1600
	Third	800	2400
	Fourth	800	3200
	Fifth	800	4000
	Sixth	800	4800
	Seventh	800	5600

The Acceptable Quality Level (AQL) columns run: 0.010, 0.015, 0.025, 0.040, 0.065, 0.10, 0.15, 0.25, 0.40, 0.65, 1.0, 1.5, 2.5, 4.0, 6.5, 10, 15, 25, 40, 65, 100, 150, 250, 400, 650, 1000, each subdivided into Ac (acceptance number) and Re (rejection number) columns.

↓↑ = Use first sampling plan below arrow. If sample size equals or exceeds lot or batch size, do 100 percent inspection.
◇◇ = Use first sampling plan above arrow (refer to preceding page when necessary).
Ac = Acceptance number.
Re = Rejection number.
* = Use corresponding single sampling plan (or alternatively, use multiple sampling plan below, where available).
∗ = Acceptance not permitted at this sample size.

* Reproduced from MIL–STD 105D, "Sampling Procedures and Tables for Inspection by Attributes,"
Superintendent of Documents, Washington, DC, 1963.

189

TABLE 7.10 Multiple sampling plans for reduced inspection

Sample size code letter	Sample	Sample size	Cumulative sample size	Acceptable Quality Levels (reduced inspection)†																							
				0.010	0.015	0.025	0.040	0.065	0.10	0.15	0.25	0.40	0.65	1.0	1.5	2.5	4.0	6.5	10	15	25	40	65	100	150	250	400	650	1000
				Ac Re	Ac Re	Ac Re	Ac Re	Ac Re	Ac Re	Ac Re	Ac Re	Ac Re	Ac Re	Ac Re	Ac Re	Ac Re	Ac Re	Ac Re	Ac Re	Ac Re	Ac Re	Ac Re	Ac Re	Ac Re	Ac Re	Ac Re	Ac Re	Ac Re	Ac Re

(Code letters A, B, C, D, E carry only arrows and asterisks — no numerical sampling plans shown.)

Code letter F — sample size 2

Sample	Sample size	Cumulative	2.5 Ac Re	4.0 Ac Re	6.5 Ac Re	10 Ac Re	15 Ac Re	25 Ac Re	40 Ac Re	65 Ac Re
First	2	2	** 2	** 2	** 3	** 3	** 4	** 4	** 5	** 6
Second	2	4	0 2	0 3	0 3	0 4	0 5	1 6	1 7	3 9
Third	2	6	0 2	0 3	0 4	0 5	1 6	2 8	3 9	6 12
Fourth	2	8	0 3	1 4	1 5	2 6	3 7	5 10	6 12	8 15
Fifth	2	10	0 3	1 4	2 6	3 7	4 8	7 11	9 13	11 20
Sixth	2	12	1 3	1 5	3 6	4 7	5 9	9 12	12 14	14 22
Seventh	2	14	1 3	2 5	4 7	6 8	6 10	9 14	13 17	18 22

Code letter G — sample size 3

Sample	Sample size	Cumulative	1.5 Ac Re	2.5 Ac Re	4.0 Ac Re	6.5 Ac Re	10 Ac Re	15 Ac Re	25 Ac Re	40 Ac Re	65 Ac Re
First	3	3	** 2	** 2	** 3	** 3	** 4	** 4	0 5	0 6	0 9
Second	3	6	0 2	0 3	0 3	0 4	0 5	1 6	3 7	3 9	3 12
Third	3	9	0 2	0 3	0 4	0 5	1 6	2 8	5 9	6 12	6 15
Fourth	3	12	0 3	1 4	1 5	2 6	3 7	5 10	8 11	9 15	11 17
Fifth	3	15	0 3	1 4	2 6	3 7	4 8	7 11	11 13	12 17	14 20
Sixth	3	18	1 3	1 5	3 6	4 7	6 9	9 12	14 15	15 18	18 22
Seventh	3	21	1 3	2 5	4 7	6 8	6 10	9 14	13 17	18 22	

Code letter H — sample size 5

Sample	Sample size	Cumulative	1.0 Ac Re	1.5 Ac Re	2.5 Ac Re	4.0 Ac Re	6.5 Ac Re	10 Ac Re	15 Ac Re	25 Ac Re	40 Ac Re
First	5	5	** 2	** 2	** 3	** 3	** 4	** 4	0 5	0 6	0 9
Second	5	10	0 2	0 3	0 3	0 4	0 5	1 6	3 7	3 9	3 12
Third	5	15	0 2	0 3	0 4	0 5	1 6	2 8	5 9	6 12	6 15
Fourth	5	20	0 3	1 4	1 5	2 6	3 7	5 10	8 11	9 15	11 17
Fifth	5	25	0 3	1 4	2 6	3 7	4 8	7 11	11 13	12 17	14 20
Sixth	5	30	1 3	1 5	3 6	4 7	6 9	9 12	14 15	15 18	18 22
Seventh	5	35	1 3	2 5	4 7	6 8	6 10	9 14	13 17	18 22	

Code letter J — sample size 8

Sample	Sample size	Cumulative	0.65 Ac Re	1.0 Ac Re	1.5 Ac Re	2.5 Ac Re	4.0 Ac Re	6.5 Ac Re	10 Ac Re	15 Ac Re	25 Ac Re
First	8	8	** 2	** 2	** 3	** 3	** 4	** 4	0 5	0 6	0 9
Second	8	16	0 2	0 3	0 3	0 4	0 5	1 6	3 7	3 9	3 12
Third	8	24	0 2	0 3	0 4	0 5	1 6	2 8	5 9	6 12	6 15
Fourth	8	32	0 3	1 4	1 5	2 6	3 7	5 10	8 11	9 15	11 17
Fifth	8	40	0 3	1 4	2 6	3 7	4 8	7 11	11 13	12 17	14 20
Sixth	8	48	1 3	1 5	3 6	4 7	6 9	9 12	14 15	15 18	18 22
Seventh	8	56	1 3	2 5	4 7	6 8	6 10	9 14	13 17	18 22	

Code letter K — sample size 13

Sample	Sample size	Cumulative	0.40 Ac Re	0.65 Ac Re	1.0 Ac Re	1.5 Ac Re	2.5 Ac Re	4.0 Ac Re	6.5 Ac Re	10 Ac Re	15 Ac Re
First	13	13	** 2	** 2	** 3	** 3	** 4	** 4	0 5	0 6	0 9
Second	13	26	0 2	0 3	0 3	0 4	0 5	1 6	3 7	3 9	3 12
Third	13	39	0 2	0 3	0 4	0 5	1 6	2 8	5 9	6 12	6 15
Fourth	13	52	0 3	1 4	1 5	2 6	3 7	5 10	8 11	9 15	11 17
Fifth	13	65	0 3	1 4	2 6	3 7	4 8	7 11	11 13	12 17	14 20
Sixth	13	78	1 3	1 5	3 6	4 7	6 9	9 12	14 15	15 18	18 22
Seventh	13	91	1 3	2 5	4 7	6 8	6 10	9 14	13 17	18 22	

◇ = Use first sampling plan below arrow (refer to continuation of table on following page, when necessary). If sample size equals or exceeds lot or batch size, do 100 percent inspection
△ = Use first sampling plan above arrow
Ac = Acceptance number
Re = Rejection number
* = Use corresponding single sampling plan (or alternatively, use multiple sampling plan below, where available)
: = Use corresponding double sampling plan (or alternatively, use multiple sampling plan below, where available)
** = Acceptance not permitted at this sample size
† = If, after the final sample, the acceptance number has been exceeded, but the rejection number has not been reached, accept the lot but reinstate normal inspection (see 10.1.4)

TABLE 7.10 Continued

Acceptable Quality Levels (reduced inspection)[†]

Sample size code letter	Sample	Sample size	Cumulative sample size	0.010 Ac Re	0.015 Ac Re	0.025 Ac Re	0.040 Ac Re	0.065 Ac Re	0.10 Ac Re	0.15 Ac Re	0.25 Ac Re	0.40 Ac Re	0.65 Ac Re	1.0 Ac Re	1.5 Ac Re	2.5 Ac Re	4.0 Ac Re	6.5 Ac Re	10 Ac Re	15 Ac Re	25 Ac Re	40 Ac Re	65 Ac Re	100 Ac Re	150 Ac Re	250 Ac Re	400 Ac Re	650 Ac Re	1000 Ac Re
L	First	20	20																										
	Second	20	40																										
	Third	20	60																										
	Fourth	20	80																										
	Fifth	20	100																										
	Sixth	20	120																										
	Seventh	20	140																										
M	First	32	32																										
	Second	32	64																										
	Third	32	96																										
	Fourth	32	128																										
	Fifth	32	160																										
	Sixth	32	192																										
	Seventh	32	224																										
N	First	50	50																										
	Second	50	100																										
	Third	50	150																										
	Fourth	50	200																										
	Fifth	50	250																										
	Sixth	50	300																										
	Seventh	50	350																										
P	First	80	80																										
	Second	80	160																										
	Third	80	240																										
	Fourth	80	320																										
	Fifth	80	400																										
	Sixth	80	480																										
	Seventh	80	560																										
Q	First	125	125																										
	Second	125	250																										
	Third	125	375																										
	Fourth	125	500																										
	Fifth	125	625																										
	Sixth	125	750																										
	Seventh	125	875																										
R	First	200	200																										
	Second	200	400																										
	Third	200	600																										
	Fourth	200	800																										
	Fifth	200	1000																										
	Sixth	200	1200																										
	Seventh	200	1400																										

⇩ = Use first sampling plan below arrow. If sample size equals or exceeds lot or batch size do 100 percent inspection
⇧ = Use first sampling plan above arrow (refer to preceding page when necessary)
Ac = Acceptance number.
Re = Rejection number.
* = Acceptance not permitted at this sample size
† = If, after the final sample, the acceptance number has been exceeded, but the rejection number has not been reached, accept the lot, but reinstate normal inspection (see 10.1.4)

* Reproduced from MIL-STD 105D, "Sampling Procedures and Tables for Inspection by Attributes," Superintendent of Documents, Washington, DC, 1963.

SUMMARY

☐ Inspection is the examination of products to determine their conformance to specifications. Inspection may be a dimensional check, physical test, performance evaluation, or visual check. The type of inspection is determined by cost or risk factors.

☐ Sampling is a statistically based technique for selecting the number of products to inspect from a population of products.

☐ Inspection is used in three operational areas in a production facility: incoming, in-process, and final inspection. Following inspection, products are accepted or rejected. If accepted, incoming products go to inventory or to production, in-process products go to the next operational area, and final products go to the customer. If rejected, products are reworked or scrapped.

☐ Planning is required in any inspection. Otherwise scarce organizational resources are wasted. Inspection planning involves the following elements: identify customer requirements, identify managerial responsibilities, identify required resources; identify quality attributes; determine frequency of inspection; define inspection methods and acceptance criteria; report inspection results.

☐ Acceptance sampling is the technical term describing sampling. The goal of acceptance sampling is to progress eventually to prevention, such as statistical process control (SPC). In acceptance sampling, a representative sample is pulled from a population of products and an accept/reject decision of the population is made based on the results of the inspection of the sample.

☐ Two methods for selecting a sample are probability and nonprobability. Probability methods assume all the products in the population are represented in the sample. Nonprobability sampling relies on the subjective judgment of the quality analyst pulling products from the population.

☐ Three methods for selecting a sample are random, systematic, and stratified sampling. In random sampling, products are pulled randomly from the population. In systematic sampling, products are organized in a manner to obtain manageable sized samples. In stratified sampling, the population of products is divided into manageable strata.

☐ Operating characteristic (OC) curves are graphs that indicate the probability of accepting a lot for various lot fraction defectives.

☐ Two types of sampling are often used: 100% and statistical sampling. 100% sampling means that all the products in a batch are inspected. It is used to comply with government regulations, to satisfy special customer requirements, and for one-of-a-kind products.

☐ 100% sampling has problems. It is expensive, inaccurate, boring, impractical, and becomes a sorting procedure.

☐ Statistical sampling is preferred over 100% sampling because it is less expensive and more efficient, results in less handling damage, and is a strong positive signal. Statistical sampling has several drawbacks. There is a risk of accepting bad lots and rejecting good lots, planning and training are required, and sampling is still inspection and not prevention.

☐ Any type of sampling assumes the following: lots are homogeneous, samples are randomly selected, and inspectors follow prescribed procedures.

☐ MIL-STD 105D has three types of sampling plans: single, double, and multiple. A sampling plan is chosen based on costs and the required level of protection.

☐ MIL-STD 105D is the most extensively used sampling plan. It is easy to learn, effective, and adaptable to different situations. This sampling plan is also called sampling by attributes. The analyst can change the level of inspection, the sampling condition, and AQL.

KEY TERMS

accept Determination, based on information obtained from a sample, that a batch, lot, or shipment of products satisfies requirements or a service complies with procedures.

acceptable quality level (AQL) Maximum percentage of defective products in a shipment that can be considered satisfactory as a process average.

acceptance number Maximum number of defective products in a sample that will allow acceptance of the shipment of products.

acceptance sampling Sampling inspection procedure to accept or reject a shipment, or batch, of products.

consumer's risk Probability of accepting a lot that should be rejected; also called beta risk.

double sampling Selecting up to two samples and accepting or rejecting a lot based on the results of the first sample; if the results of the first sample are inconclusive, then in the second sample the entire lot is accepted or rejected.

final inspection Examination of final products; usually involves performance testing of final product or assembly.

incidental sampling Same as nonprobability sampling.

incoming material inspection Examination of products from external suppliers; products may be partially finished goods, components, or capital equipment.

in-process inspection Examination of partially processed products as they move down the production line.

inspection Examination, including measuring, visually checking, or physically testing products.

inspection level Element of the sampling plan that relates sample size to lot size; sampling plans may have normal, reduced, or tightened inspection levels.

inspection planning Following a series of steps for determining inspection requirements.

lot size Number of products in a lot, batch, or shipment.

multiple sampling Pulling samples from a population of products until a decision to accept or reject the entire population is made.

nonprobability sample Selection of products in a sample relying on the subjective judgment of the inspector; sample is not derived statistically.

100% sampling Sampling of all products in a shipment, batch, or lot.

operating characteristic (OC) curve Graph indicating the probability of accepting a lot for various percent defectives.

probability sample Sample representing all products in a population.

producer's risk Probability of rejecting a lot that should be accepted; also called alpha risk.

random sampling Process of selecting sample products so that each product has an equal chance of being selected in the sample.

reduced inspection Element of the sampling plan that allows for smaller sample sizes than used in normal inspection.

reworked Nonconforming products having been returned to the source and repaired.

sample size Number of products in the sample pulled from a lot, batch, or shipment.

sampling Statistically based method for selecting products from a population.

sampling plan Determining the number of products to be pulled from a shipment for inspection, given lot size, acceptable risk, and level of inspection.

single sampling Making an accept/reject decision based on the inspection of a single sample.

statistical sampling Sampling of a small representative selection of a shipment, batch, or lot.

stratified sampling Sampling technique to divide a lot or shipment into strata or layers.

systematic sampling Pulling products from a population based on a logical method.

tightened inspection Element of a sampling plan that allows for larger sample sizes than used in normal inspection.

QUESTIONS

1. Define sampling and inspection. How are the two related?

2. Inspection can be used in three operational areas. Discuss each.

3. Should final inspection focus on dimensional checks or performance evaluation? Discuss.

4. Following inspection, products are scrapped or reworked. Why is this expensive?

5. If inspection or sampling is not planned, scarce corporate resources may be wasted. Explain.

6. What are the elements of the inspection plan? Discuss each.

7. Frequency of inspection is based on what factors?

8. What is acceptance sampling?

9. Is it accurate to say that a sample is necessarily representative of the population?

10. What is the difference between probability and nonprobability methods of sampling?

11. What are the three methods of pulling a sample? Discuss each.

12. Give an example of a sampling procedure.

13. What is an operating characteristic (OC) curve?

14. What is the difference between 100% and statistical sampling and what are the relative advantages of each?

15. When should 100% sampling be used? Give an example of its use.

16. What type of signal does sampling send to an organization or to the person responsible for quality?

17. What is the difference between producer's and consumer's risk?

18. Discuss each of the assumptions of sampling.

19. What are three types of sampling plans? Discuss each.

20. What information must be known before MIL-STD 105D can be used?

PROBLEMS

1. A single sampling plan uses a sample size of 10 and an acceptance number of 1. Using the Poisson probability distribution equation discussed in Chapter 4, what is the probability of accepting a lot of 100 articles that are 1%, 5%, and 10% defective?

2. Using MIL-STD 105D, single sampling, level II, AQL = 10%, and a lot size of 1000, what are the acceptance criteria under:
   a. Normal inspection?
   b. Tightened inspection?
   c. Reduced inspection?

3. Using MIL-STD 105D, single sampling, level II, AQL = 2.5%, and a lot size of 100, what are the acceptance criteria under:
   a. Normal inspection?
   b. Tightened inspection?
   c. Reduced inspection?

4. Using MIL-STD 105D, double sampling, level II, AQL = 10%, and a lot size of 200, what are the acceptance criteria for both samplings under:
   a. Normal inspection?
   b. Tightened inspection?
   c. Reduced inspection?

5. Using MIL-STD 105D, double sampling, level II, AQL = 2.5%, and a lot size of 100, what are the acceptance criteria under:

    a. Normal inspection?

    b. Tightened inspection?

    c. Reduced inspection?

6. Construct an OC curve for a single sampling plan with the following data: $N = 1000$, $n = 100$, $c = 4$. Use six points.

7. Construct an OC curve for a single sampling plan with the following data: $N = 500$, $n = 50$, $c = 3$. Use six points.

8

Auditing

Organizations rely on audits to monitor and ensure that internal quality controls are in place and are working effectively. A company-wide quality management (CWQM) program will serve to instill and reinforce goals, policies, objectives, and measurement systems in order to achieve quality.* Auditing is a major tool to monitor and evaluate the effectiveness of the CWQM program, processes, products, and people. These are called the 4 Ps and are the essential elements of CWQM.†

Almost every type of organization uses **audits.** Audits are used extensively in the public and private sectors, military organizations, schools, and the federal government. Regardless of the specific purpose of an audit, audits have a common structure and purpose and similar key players. This chapter addresses the general nature of audits.

WHAT IS AUDITING?

Quality auditing is the process of accumulating and evaluating quantifiable information. This definition of auditing is similar to that for financial auditing. Both financial and operational auditing monitor internal control systems. The former

---

* G. Hutchins, "Procurement Quality: The Competitive Edge," (QPE, 1989), All Rights Reserved.
† "Program, processes, products, and people" and "4 Ps" are service marks of Quality Plus Engineering. All rights reserved.

checks financial and cost systems, the latter the quality program, processes, products, and people. The audit is conducted by independent and skilled persons. The purpose of an audit is to determine and report upon the degree of correspondence between quantifiable information and established criteria. The degree of correspondence can range from 100% positive correlation to no correlation at all. In most cases the correlation is somewhat less than 100%. The auditor then must assess the significance of the deviation as well as the amount of deviation from the established criteria. Quantifiable information is obtained during the assessment, or audit. Established criteria can be engineering specifications, drawings, industry standards, company policies, or department procedures.*

In summary, the definition of quality auditing encompasses certain key phrases and concepts, which you will see repeated and explained throughout this chapter, specifically:

- ☐ Auditing is the process of accumulating and evaluating quantifiable information.
- ☐ Information is gathered for the CWQM program, processes, products, and people.
- ☐ Auditing evaluates the degree of correspondence between quantifiable information and established criteria.
- ☐ Audits are conducted by skilled and competent persons.

Accurate and Reliable Information

Quality auditors strive to obtain timely, accurate, reliable, unbiased, and complete information. Unreliable information can cause inefficient use of resources, which will result in deteriorating quality.

A common example may help to illustrate this point. Manufacturing companies are slowly eliminating incoming material inspection. They expect suppliers to provide defect-free material, just in time to be used in the manufacturing facility. The manufacturing company now relies on the supplier's process controls to maintain and improve product quality. The company periodically audits the supplier to ensure that process controls are being used to monitor and improve quality.

The need for reliable, accurate, and timely quality information is assuming more importance because of global marketplace competitiveness, pressures on organizations, and the availability of increasing amounts of information.

Global Marketplace Competitiveness. The global marketplace is forcing service and manufacturing organizations to focus on cost-competitive quality. To survive, service organizations must deliver courteous, reliable, and efficient services. Manufacturing organizations must offer reliable, cost-competitive, and defect-free products. To produce these products and deliver these services, cus-

---

* This definition of auditing has similar components as the definition of financial auditing found in A. Ahrens and J. Loebbecke, *Auditing: An Integrated Approach* (Prentice Hall, Englewood Cliffs, NJ, 1984).

tomer needs, wants, and expectations must be communicated accurately and quickly to everyone in the organization as well as to the organization's suppliers.

Pressures on Organizations. Management needs reliable, accurate, and timely information in order to make effective decisions. Unfortunately this information may be provided by a group who aims to please, and thus biases information in its favor. At worst, information may be purposely falsified. Whatever the reason, whether honest optimism, intentional manipulation, or negligence, the result is the same. The decision maker is receiving misstated information, which results in poor decisions that can lead to deteriorating product quality and eventual customer dissatisfaction.

Availability of Increasing Amounts of Information. The amount of available information is increasing exponentially, as evidenced by computers performing 100% inspection or on-line statistical process control (SPC). Such a volume of information increases the likelihood that improperly recorded or incorrect information is inputted into the system. Minor errors eventually compound into critical errors.

Audits and Surveys

Terms such as survey, assessment, investigation, or evaluation are sometimes used in place of auditing. In this text they are synonymous. These terms refer to an independent and knowledgeable person or persons who obtain information, evaluate it, derive conclusions, and make recommendations.

Sometimes the term audit is specifically differentiated from **survey.** A quality audit is a more formal, extensive, and intensive investigation and evaluation than a survey. Also, an audit is authorized by top management to investigate company-wide problems. A survey is a less extensive examination and is usually authorized by lower management to assess an operational problem, determine the effectiveness of corrective action, or evaluate a supplier.

AUDITOR, CLIENT, AND AUDITEE

Key Audit Players

There are three main parties in any audit: auditor, client, and auditee. The **auditor** is the person or persons conducting the audit; the **client** is the person or group requesting the audit; and the **auditee** is the person being audited.* In this section we examine each of these.

---

* "Generic Guidelines for Auditing Quality Systems," ANSI/ASQC Q1-1986, Am. Soc. for Quality Control, Milwaukee, WI, 1986.

Auditor

An audit can be conducted by one person or by a team of professionals. The size of the audit team depends on the scope of the audit, the size of the organization being audited, requirements of the client, risks involved, and costs. A program audit of a large organization is usually more costly and expensive than an evaluation of a product failure. However, if the product failure was the landing gear of an aircraft, then the assessment team would be large.

Large organizations often have internal audit groups responsible for financial and quality auditing. The financial audit group is headed by an officer of the organization and results are reported to the executive committee. The operations or quality auditing group may report to the financial group, the auditing organization, or the head of a business unit. It investigates and ensures that policies and procedures are being followed, objectives met, and results of corrective action communicated.

Staff auditors should be perceived as being independent of the person being investigated, so that there is no possibility of abuse. Unfortunately auditors reporting to the head of a business unit are sometimes perceived as ''snitches'' or ''police officers.'' Employees may think that the boss is using auditors to monitor efficiency and attitudes. This erodes the belief that everyone is responsible for his or her own quality.

Levels of Auditors. An audit of a large supplier may require a team consisting of a lead auditor, two senior auditors, and up to a dozen staff auditors. A lead auditor can lead a team to investigate major problems or evaluate CWQM programs. The lead auditor manages the audit and reports conclusions to the client. He or she selects members of the audit team, establishes audit objectives, plans audit implementation, and represents the team in client negotiations.

Senior auditors are responsible for organizing the collection, examination, and assessment of audit information. In a large audit, senior auditors are responsible for evaluating a functional area. For example, a senior auditor may have two staff auditors evaluating manufacturing quality, including process control, inspection equipment, material handling, and documentation. Staff auditors collect the evidence or information.

An auditor must have sufficient skills to perform an audit and to ensure audit effectiveness and reliability. Audit skills are developed through on-the-job experience and training under the tutelage of a senior auditor. As experience and skills are gained, a junior auditor evolves into a senior and then a lead auditor.

Client

The client is the person, department, or group requesting the audit. Another way to view this relationship is to say the client is the auditor's customer. The client usually defines the audit's scope, milestones, time lines, and standards. The scope states in general terms what work has to be performed. The milestones are the major objectives that the audit should accomplish. The time lines are the dates by which the objectives should be reached. The standards, procedures, policies, and specifications are used as benchmarks to determine the degree of correspondence or compliance between audited information and a standard.

The frequency of an audit is based on the type of audit, the cost, the critical nature of the product, customer requirements, and emergency requirements. Since program audits are more extensive, they require additional personnel, resources, and time. Program audits are conducted periodically based on the needs of the client. Supplier audits of critical components are performed at least yearly by the customer's purchasing or quality assurance departments.

Emergency situations require immediate attention and assessment. These assessments focus on finding the root cause of a problem. For example, any time there is a major oil or gas accident, the Department of Transportation dispatches a team to investigate.

Audits are initiated upon a request from a client, customer, senior management, or department. The requester must have the authority and ''need to know'' to request the audit. An audit request is sometimes sent to a higher authority or an independent board for approval. External audits of a supplier, for example, can be conducted by the client department.

A **material review board (MRB)** or an **engineering control board (ECB)** may also authorize an audit. The MRB is a group composed of department managers or representatives who decide on the acceptance, disposition, and use of material. The ECB is a group composed of personnel from manufacturing, manufacturing engineering, engineering design, purchasing, and quality assurance. It evaluates new designs, reviews modified designs, approves specifications, and comments on procedures and standards that can affect product reliability, performance, or marketability.

Auditee

The auditee can be the person, function or area being audited. An auditee has several responsibilities to facilitate the progress of the audit: cooperate and assist with the audit, provide adequate facilities and necessary equipment to complete the audit, review audit recommendations and conclusions, and implement any required corrective action.

An auditor collects data and information that are used to reach conclusions or make recommendations. The information may be buried in computer files, unorganized, or consist of handwritten notes. The auditee should appoint a staff person to facilitate obtaining the information. Department or area supervisors have been known to be possessive and turf conscious of any intruders. This attitude can forestall an audit, creating unnecessary tension between all parties involved.

The auditor or team requires facilities and equipment to complete the audit, such as space to collect and analyze information. If a product is being tested, personnel and equipment are required to test one or more products.

At the conclusion of the audit, the auditee has the opporunity to review the auditor's report and, if necessary, offer evidence to rebut perceived unfair, unreasonable, or inconclusive reports.

SPOTLIGHT

The auditee is in an interesting position. Nobody enjoys being audited. The appearance of the auditor and the resulting report card can engender fear. It means having an outsider visit, monitor, and evaluate one's operation.

The auditee has no choice in being audited. Being subjected to an audit is a normal business event. For suppliers it is a condition of doing business with certain customers. For example, a customer may audit a supplier to ensure that control systems are operating effectively; or federal or state government may audit industrial firms to check compliance with environmental regulations. Internal departments are audited to check compliance with management policies.

If a problem of nonconformance or noncompliance has been discovered, the report will recommend some action to correct the problem. The auditee may be responsible for implementing the action.

ETHICS AND AUDITOR REQUIREMENTS

It is essential that the decision maker and information users have confidence in the audit. If the users of the audit are not confident in its conclusions or recommendations, management decisions may be impaired or compromised. The auditor must have high ethical standards to conduct audits in a professional manner, obtain the required information, and make reliable assessments.

Often organizations, professional agencies, and government agencies have **codes of ethics.** Such codes state ideal types of behavior as well as specific behaviors deemed inappropriate by auditors. Furthermore, professional organizations may prescribe minimum education, background, and experience requirements before an auditor is hired to perform a particular job.

Common Elements of Most Codes of Ethics

Certain principles are found in many quality codes of ethics. The principles are distilled in the following key concepts, which are discussed briefly:

- ☐ Objectivity and independence
- ☐ Confidentiality
- ☐ Competence
- ☐ Knowledge and training
- ☐ Interpersonal skills
- ☐ Responsibility to client and organization

Objectivity and Independence. Any conflict of interest, or even the appearance of such, should be avoided. To accept a financial favor from an auditee is an actual conflict of interest. The appearance of a conflict of interest is accepting small favors from the auditee, such as presents or trips.

The auditor should be impartial and objective. Objectivity can be compromised if an auditor befriends or subordinates to the party or area being audited. If conflicts of interest or interference arise during an audit, they should be reported to top management so that the audit's impartiality will not be prejudiced or questioned.

Confidentiality. Confidentiality is essential of all auditors. Auditors often deal with proprietary and privileged information. This information can harm people or, in an extreme, provide a competitor with sensitive information that enhances market position.

In military organizations, or with defense contractors, information is often classified and an unintentional slip can lead to national security conflicts. Military supplier audits sometimes require a security clearance and audit results have a limited distribution to those with a "need to know."

Competence. **Competence** is the ability of an auditor to conduct and complete an audit according to professional standards and an organization's policies. Competence is interpreted differently depending on the perspective of the evaluator. From the client's perspective, competence is the ability to obtain required information, reach reliable conclusions, make accurate assessments, and complete the audit on time and within budget. From the auditee's perspective, competence is the ability to present a balanced and accurate assessment.

Knowledge and Training. Competence is established through technical qualifications, a knowledge of the auditee's operations, and the ability to exercise sound judgment. The auditor or team must have the appropriate qualifications and professionalism to perform an audit. Qualifications are based on skills, knowledge, and training obtained at school and on the job. Professionalism is based on competence, ethics, and integrity to perform the audit impartially and independently and is obtained through on the job experience.

Depending on the type of audit, the auditor needs a basic knowledge of the disciplines of quality, manufacturing, engineering, purchasing, and inventory control. In addition, an extensive audit of a large organization may require specialists with specific knowledge of the industry, processes, or products. In a regulated industry, auditors should be familiar with federal and industry codes, standards, regulations, and guides.

For example, an auditor evaluating a natural gas pipeline for corrosion should know U.S. Department of Transportation (DOT) natural gas regulations as well as industry standards, such as those of the American Gas Association (AGA) and the National Association of Corrosion Engineers (NACE).

Interpersonal Skills. The auditor must have interpersonal skills to obtain the necessary cooperation of the auditee. A thorough knowledge of an auditee's operations is useless without an equivalent ability to get along with people and solicit the information. The auditor needs patience, persistence, and the ability to take the heat and probe until, any "reasonable doubt" has been eliminated. Once

SPOTLIGHT

The following is a code of ethics of a professional association. The code has elements common to many codes of ethics.

☐ To promote the advancement of the association and its beliefs.
☐ To maintain sound business practices, engender high standards of professional conduct, and follow this code of ethics.
☐ To maintain confidentiality of any information gained in an organization and to refrain from using such information in an unethical manner.
☐ To develop abilities, knowledge, and information through constant study.
☐ To maintain high personal standards of moral responsibility, character, and integrity.
☐ To never exploit membership, company, or profession.
☐ To refrain from using the name of the organization for personal gain.

cooperation is assured, the auditor needs investigative and analytical skills to obtain the necessary information.

Responsibility to Client and Organization. The auditor has responsibilities to the client as well as to the organization he or she represents. An obvious responsibility is not to accept a commission or bribe so that audit judgments cannot be questioned and that not even an appearance of doubt will arise.

Another responsibility is to have an unbiased perspective when conducting an audit, which comes down to confidence. The client should have confidence in the results of the audit. Independence is essential in fact and in appearance. Sometimes the appearance of a conflict of interest can be as damaging as an actual conflict.

TYPES OF QUALITY AUDITS

Audits are differentiated by type, size, nature of investigation, or degree of correspondence between the audited information and standards. In this section we explain four basic types of audits: program, process, product, and people audits.*

Program

A **program audit** evaluates the company-wide quality management (CWQM) program. It ensures that a quality program is in place and is operating as policies and procedures prescribe. A program audit is an overall assessment of an organization's quality structure. If required, it may also assess processes and a product. A program audit may also encompass process audits of every functional area in the

---

* G. Hutchins, ''Supplier Auditing,'' Seminar Notes, 1989.

SPOTLIGHT

The following issues are addressed in a quality program audit:

☐ Is there a company-wide quality program?
☐ Does top management support the program?
☐ Are quality culture and philosophies pervasive throughout the organization?
☐ Are quality policies, plans, objectives, accountabilities, and measurements utilized and updated as required?
☐ Is the quality program based on defect prevention?
☐ Is personnel trained thoroughly in quality techniques and philosophies?

organization, evaluate specific product dimensions, and evaluate worker effectiveness.

A program audit specifically evaluates the management structure, culture, documentation, policies, objectives, controls, and measurement system of an organization. A CWQM program must be supported by top management in order to succeed. The CEO should take an active role in supporting and directing the program. The CEO is the main person in the organization who can establish a culture of quality. Policies explain the organization's quality expectations. Objectives state goals for the organization and then for each business unit. Controls are in place to ensure that objectives are met and, if not, that the system can adjust to

SPOTLIGHT

The following typical questions are asked when auditing quality assurance, purchasing, and manufacturing:

Quality Assurance

☐ Is quality assurance independent and does it have direct access to high levels of company management?
☐ Does quality assurance maintain and update a quality manual?
☐ Do corrective action procedures eliminate root cause problems?

Purchasing

☐ Do suppliers have a company-wide quality program?
☐ Are suppliers evaluated and monitored for quality?
☐ Is corrective action prompt and effective?

Manufacturing

☐ Is statistical process control used?
☐ Is personnel trained?
☐ Is measuring and test equipment available, adequate, and calibrated?

meet the objectives. Every organizational activity with a quality element is measured.

Processes

Process audits evaluate internal and external functional areas, including marketing, engineering, manufacturing, purchasing, distribution, and accounting. These functional areas process resources, be it information, paper, money, raw material, or process material.

Process audits assess whether quality controls, checks and balances, and a documentation trail are in place in the functional areas of the organization. If processes are being monitored, feedback loops should be in place to correct any discrepancies, and audits should be performed periodically. Then there is a level of assurance that the outputs of the processing products or services comply with specifications and satisfy the customer.

Products

Products can be an intangible service or tangible products. More often, a product combines the two. Thus tangible **product audits** assess whether products conform to dimensional, performance, or maintainability standards. For example, a product audit may evaluate basic product characteristics such as packaging, warranty provisions, storage, and handling.

The customer may be satisfied with a product but dissatisfied with its delivery or service. Regardless of the tangible quality of the product, if the service is poor, the customer will be dissatisfied. This will affect future sales.

People

People, or personnel, audits are difficult to implement because they evaluate the performance and effectiveness of people. W. Edwards Deming and other quality authorities believe that management should identify and eliminate organizational impediments that hinder people from contributing up to their potential. Audits can be used to find these obstructions. Audits can also be used to determine how well the needs of internal customers are being satisfied. Properly developed and imple-

SPOTLIGHT

Product quality questions focus on conformance to specifications, product reliability, and product safety:

- ☐ Are customers satisfied with the products?
- ☐ Do products conform to specifications?
- ☐ Are products safe, reliable, and maintainable?

mented, people audits provide significant quality information, which the organization can use to guide and reward performance.

AUDITING METHODOLOGY

While many types of organizations use audits, and audits have different purposes, the steps involved in these audits are similar. They involve planning, implementation, and reporting.

Planning

All types of audits should be planned prior to implementation. An audit plan is a map for conducting the audit. The audit plan serves several important purposes. First, it is used to determine the purpose, scope, and time lines of the audit. Second, it is used to verify the existence of a problem. Is the problem real and does it require an audit? Third, a review identifies the resources needed to complete the audit.

Audit planning is performed by a team composed of auditor, client, and auditee. This brings together all the key players in the audit. The auditor identifies resources needed to complete the audit. The client explains the purpose and scope of the audit. The auditee offers resources required to gather information and complete the audit.

If an audit is not planned properly, the results may not be accurate. The information must be obtained from the right sources, distilled, analyzed, and presented to the auditee in an understandable form. Proper planning also minimizes the potential of future problems. Extensive and large quality system audits can continue for the duration of a project. For example, the Nuclear Regulatory Commission (NRC) requires nuclear power plants to comply with extensive codes and standards while under construction and during operation. Resident audit teams check compliance on a daily basis.

An important element of planning is notifying the auditee's management and the auditee that the audit is to take place. This is more than a courtesy gesture. The auditee has to prepare for the audit by having personnel, equipment, and records available. Under certain circumstances audits may also be unannounced. An unannounced audit might be conducted to check operational readiness or compliance with specifications. Military readiness audits verify that operational units can be mobilized quickly in a state of emergency.

Implementation

Sufficient relevant information has to be gathered to permit a valid and reliable examination. Information is considered relevant and sufficient if another qualified and independent auditor would reach similar recommendations and conclusions based on the same analysis of the same information.

The audit implementation stage consists of various techniques to assess the degree of correspondence between quantifiable data and established criteria.

Quantifiable data can be objective physical evidence such as test, measurement, or inspection results; or it can be subjective evidence or oral information retrieved from inspectors or direct observation. Auditors can obtain these data through product testing, sampling, observing operations, measuring critical characteristics, and interviewing people. The appropriate method of obtaining data is based on the risk of obtaining a bad judgment, cost, time, and the resources available to perform the audit. The auditor must know the relative importance of each item of evidence and determine how much credence to give it. Facts and objective evidence gathered by the auditor are more reliable than subjective data obtained from interviews. Any subjective data should be corroborated whenever possible. If a series of deficiencies or nonconformances are discovered, these are prioritized by degree of importance, ease of solution, or cost-effectiveness.

Audits may conclude that a situation is satisfactory, that is, there are no significant deficiencies, flaws, nonconformances, or mistakes to warrant further action. Or an audit may conclude that there are significant deficiencies, which can be operational inefficiencies, nonconformances, defectives, defects, or noncompliance with regulations.

If a deficiency is found, corrective action is initiated. Correcting a cause may not eliminate the problem. The correction may eliminate a symptom at one point in time, not the root cause. For example, a boiler does not supply hot water. The quick fix is to remove the immediate cause, which in this case, is to clean the boiler tubes and replace the water filter thus improving heat transfer. But the root cause is dirty water and the lack of a preventive maintenance (PM) program. Even though the filter cleans the water, the PM program would ensure that filters are replaced periodically and the tubes checked regularly for rust and contamination.

Reporting

The auditor, or team, finally prepares a report and debriefs the auditee and the client. The purpose of the postaudit debriefing with the auditee is to present initial observations, conclusions, and recommendations. At this point the auditee has the opportunity to rebut the conclusions and recommendations. If there is disagreement, this is the time for the auditee to present offsetting evidence.

An audit presents a picture of a product or process at one point in time. So an audit may recommend corrective action and a subsequent audit to review the effectiveness of the corrective action. Corrective action is implemented to correct any deficiency or nonconformance so that it will not recur. Any corrective action not only focuses on solving the immediate cause or symptom of the discrepancy but, more importantly, focuses on eliminating the recurrence of the discrepancy.

SPOTLIGHT

The following is an abbreviated supplier quality audit. The audit form is specially tailored for auditing process industries. Quality audits can be developed around specific national, industry, process, or product requirements. This audit evaluates six supplier areas: organization; handling, packaging, and storage control; laboratory; raw material control; in-process control; and final product control.

Product quality is verified by the existence and effectiveness of quality control and assurance systems. The assumption is that if sufficient systems are in place and working, there is a level of assurance about the quality of the products being produced by these systems.

This type of audit is relatively easy to conduct. The auditor has a checklist of questions. Each question has four responses, Yes, PYes (partially yes), No, and N/A (not applicable). Any time there is a No or PYes response, the auditor notes the negative response at the end of the audit and makes a recommendation for subsequent action. The auditor may recommend that corrective action be pursued to eliminate the cause of the problem, or the auditor may waive this noncompliance because he or she believes that the question is not relevant or important to improving quality.

Every audit section is assigned a percentage and all sections combined should total 100%. In this example an internal group composed of representatives from engineering, purchasing, manufacturing, and accounting determined that laboratory and in-process control were more important in determining and controlling quality than the other factors. Laboratory and in-process control were therefore each assigned 20%, while the other four factors were assigned 15% each. Each question in a section is assigned a point value. At the end of the audit, the auditee has the opportunity to review and comment on the auditor's recommendations and conclusions.

Organization (15%)

1. Is there a corporate quality improvement goal?
   Yes   PYes   No   N/A
2. Is there a corporate quality policy statement?
   Yes   PYes   No   N/A
3. Is there a corporate quality management group?
   Yes   PYes   No   N/A
4. Is the quality group separate from and equal to other operating groups?
   Yes   PYes   No   N/A
5. Is there a quality manual?
   Yes   PYes   No   N/A
6. Does the quality manual address:
   a. Organization?   Yes   PYes   No   N/A
   b. Statistical process control?   Yes   PYes   No   N/A
   c. Raw material control?   Yes   PYes   No   N/A
   d. In-process material control?   Yes   PYes   No   N/A
   e. Finished product control?   Yes   PYes   No   N/A
   f. Internal audits?   Yes   PYes   No   N/A
   g. Corrective action?   Yes   PYes   No   N/A
7. Is there a quality training program?
   Yes   PYes   No   N/A

Handling, Packaging, and Storage Control (15%)

1. Is there a packaging control system?
                    Yes   PYes   No   N/A
2. Are there packaging instructions?
                    Yes   PYes   No   N/A
3. Is there a system to ensure proper labeling and identification?
                    Yes   PYes   No   N/A
4. Is there a storage control system?
                    Yes   PYes   No   N/A
5. Are there storage instructions?
                    Yes   PYes   No   N/A
6. Is there a system to ensure proper handling?
                    Yes   PYes   No   N/A
7. Is there a process to document, analyze, and respond to customer complaints?
                    Yes   PYes   No   N/A
8. Are customer complaints used to improve the process/product?
                    Yes   PYes   No   N/A

Laboratory (20%)

1. Does the quality manual describe all laboratory procedures?
                    Yes   PYes   No   N/A
2. Does the lab manual address final product specifications?
                    Yes   PYes   No   N/A
3. Does lab personnel understand quality requirements?
                    Yes   PYes   No   N/A
4. Does the sampling plan indicate the names of samples to be taken, the times, amounts, and types of analyses performed?
                    Yes   PYes   No   N/A
5. Are lab results complete, understandable, and accurate?
                    Yes   PYes   No   N/A
6. Are records complete and accurate?
                    Yes   PYes   No   N/A
7. Are instruments and gaging instructions available?
                    Yes   PYes   No   N/A
8. Are instruments calibrated regularly and traceable to National Institute of Standards and Technology (NIST)?
                    Yes   PYes   No   N/A
9. Are instruments periodically maintained?
                    Yes   PYes   No   N/A
10. Are instruments stored properly?
                    Yes   PYes   No   N/A
11. Are SPC control charts used?
                    Yes   PYes   No   N/A

Raw Material Control (15%)

1. Does a raw material quality control program exist?

   Yes   PYes   No   N/A

2. Are (auditee's) suppliers required to use SPC techniques?

   Yes   PYes   No   N/A

3. Is there a supplier rating system?

   Yes   PYes   No   N/A

4. Are (auditee's) suppliers audited?

   Yes   PYes   No   N/A

5. Are specifications available for all incoming raw materials?

   Yes   PYes   No   N/A

6. Are certain shipments tested?

   Yes   PYes   No   N/A

7. Does documentation define responsibility for acceptance/rejection of incoming raw material?

   Yes   PYes   No   N/A

8. Does documentation define criteria for acceptance/rejection of incoming raw material?

   Yes   PYes   No   N/A

9. Are rejected shipments identified and effectively segregated?

   Yes   PYes   No   N/A

10. Are (auditee's) suppliers notified of quality problems?

    Yes   PYes   No   N/A

11. Is corrective action effective?

    Yes   PYes   No   N/A

In-Process Control (20%)

1. Are all critical processes monitored and controlled?

   Yes   PYes   No   N/A

2. Does the quality manual indicate where quality is monitored?

   Yes   PYes   No   N/A

3. Does it indicate how quality is monitored?

   Yes   PYes   No   N/A

4. Does it indicate who monitors quality?

   Yes   PYes   No   N/A

5. Do process control instructions exist?

   Yes   PYes   No   N/A

6. Is responsibility to approve process changes clearly defined?

   Yes   PYes   No   N/A

7. Is there a procedure for giving customers advance notice of process changes?

   Yes   PYes   No   N/A

8. Is SPC used for process control?

   Yes   PYes   No   N/A

9. Are all critical processes
   a. In statistical control?   Yes   PYes   No   N/A
   b. Capable?   Yes   PYes   No   N/A
   c. Improved?   Yes   PYes   No   N/A
10. Are the following statistical methods used in solving problems:
   a. Pareto analysis?   Yes   PYes   No   N/A
   b. Histograms?   Yes   PYes   No   N/A
   c. Cause-and-effect analysis?   Yes   PYes   No   N/A
11. Are the following readily available or posted:
   a. Specifications, standards?   Yes   PYes   No   N/A
   b. SPC charts?   Yes   PYes   No   N/A
   c. Critical product parameters?   Yes   PYes   No   N/A

Final Product Control (15%)

1. Is there a formal documented specifications system?
                        Yes   PYes   No   N/A
2. Does personnel understand specifications?
                        Yes   PYes   No   N/A
3. Is there a final product quality control system and does it include:
   a. Sampling?   Yes   PYes   No   N/A
   b. Specifications?   Yes   PYes   No   N/A
   c. Nonconformances?   Yes   PYes   No   N/A
   d. Storage?   Yes   PYes   No   N/A
   e. Traceability?   Yes   PYes   No   N/A
4. Is the authority to reject nonconforming product defined and documented?
                        Yes   PYes   No   N/A
5. Are nonconforming products clearly identified and segregated?
                        Yes   PYes   No   N/A
6. Are SPC charts attached to shipments?
                        Yes   PYes   No   N/A
7. Is there a procedure to notify customer if nonconforming products are returned?
                        Yes   PYes   No   N/A

Action

Score:
Auditee's review and comments:
Auditor's recommendations and conclusions:
Corrective action required:
Corrective action due date:

AUDIT MISUSE AND ABUSE

Sometimes problems arise in auditing. An auditor can have a conflict of interest, can report audit details to an unauthorized person, or can accept moneys in exchange for biasing a report. So the tool itself should be monitored and reviewed.

Who audits the auditors? In some organizations the auditees audit the performance of the auditing organization. Sometimes the clients form an ad hoc committee to monitor the audit group. This way the audit group is kept honest and does not override its authority. The audit group should be evaluated periodically for:

- ☐ Independence
- ☐ Ethics
- ☐ Effectiveness and efficiency
- ☐ Competence
- ☐ Cost-effectiveness and benefits
- ☐ Quality of recommendations and conclusions

SUMMARY

☐ Auditing is a major tool to monitor and evaluate the CWQM program, processes, products, and people. Audits are used extensively in the public and private sectors to monitor, evaluate, and correct these elements of the CWQM system.

☐ Quality auditing is the process of accumulating and evaluating quantifiable information of the CWQM program, processes, products, and people. Audits evaluate the degree of correspondence between quantifiable information and established criteria.

☐ Audits are becoming more important because of global marketplace competitiveness, additional pressures on organizations, and ever-increasing amounts of information to interpret.

☐ The three key players in any audit are the auditor, the client, and the auditee. The auditor is the person or persons who conduct the audit. The client is the person or group that requests the audit. The auditee is the area, person, or group being audited.

☐ Audits are conducted by skilled persons with the proper education, training, and aptitudes. Being ethical in the conduct, implementation, and re-

porting of an audit is a prerequisite of any auditor. Ethics involves the following: objectivity and independence, confidentiality, competence, knowledge and training, interpersonal skills, and responsibility to client and organization.

☐ Audits are differentiated by type, nature, and size. In this chapter we discussed four types of audits: program, process, product, and people. Program audit assesses the company-wide quality management program. Audits of processes evaluate functional areas, including marketing, engineering, manufacturing, purchasing, distribution, and accounting. Audits of products assess intangible services or tangible products to ensure compliance to specifications or procedures. People audits evaluate personnel and its effectiveness. People audits are the most difficult to conduct because of the human element.

☐ Audits follow a series of sequential steps consisting of planning, implementation, and reporting results. An audit plan is a road map for conducting an audit. Once a plan has been formulated, an audit is conducted. Finally, the auditor or audit team reports the results of the audit to the client.

KEY TERMS

audit Process of accumulating and evaluating quantifiable information to determine the degree of correspondence between quantifiable information and established criteria.

auditee Program, process, product, or person being audited.

auditor Person conducting the audit.

client Person or persons requesting the audit.

code of ethics Standard of an organization or professional group governing the professional behavior of its membership.

competence Ability to conduct and complete an audit according to professional standards.

engineering control board (ECB) Group composed of department managers or representatives who decide on new designs, review modified designs, and approve specifications.

lead auditor Leader of audit team; manages and reports audit conclusions to client.

material review board (MRB) Group composed of department managers or representatives who decide on the acceptance, use, and disposition of material.

people audit Evaluation of the effectiveness of the people who perform an operation.

process audit Evaluation of a functional area of an organization, such as marketing, purchasing, or accounting.

product audit Evaluation of either a tangible product or an intangible service.

program audit Evaluation of the company-wide quality management program.

senior auditor Auditor responsible for organizing the collection, examination, and assessment of audit information.

survey Examination of a problem; less extensive than an audit.

QUESTIONS

1. Define auditing. What is the difference between an audit and a survey?

2. Why is timely, accurate, and reliable information required?

3. Discuss the factors that make the need for independent quality audits important.

4. Auditors are required to have sufficient technical proficiency and training. What are various ways these can be developed?

5. List four different types of audits and discuss each.

6. What responsibilities does an auditor have to the client?

7. What are the responsibilities of the auditee?

8. Explain the purpose for a code of ethics.

9. While conducting an audit the auditor determines that there is a conflict of interest. What should the auditor do?

10. What is the role of the lead auditor?

11. Distinguish between independence in fact and in appearance.

12. Why are people audits difficult to conduct?

13. What are the three elements of any audit? Discuss each.

14. How can audits be abused? How can abuse be deterred?

9

Reliability

Reliability is long-term quality. The topic of reliability is becoming more important due to shortened product life cycles, heightened customer expectations, the high cost of failures, evolution of complex systems, and safety considerations. Each factor is a risk that reliability engineers try to control and minimize.

Reliability is an attribute of quality that is important in the defense, aerospace, and health industries. Nuclear power plants, military systems, and mechanical hearts go through extensive reliability testing, federal approvals, and inspection before they are approved for use. In case of failure, they may have redundant features. For example, a rocket may have an extra engine in each stage to propel the vehicle if one engine should fail.

Furthermore, reliability is assuming greater importance because of the risk of litigation and high jury awards. Companies have been held liable for accidents 20 or more years after a product was designed and manufactured. Companies have also been held liable for products that were misused or abused because there were no warning labels.

IMPORTANCE OF RELIABILITY

Most people's understanding of reliability centers on products, such as the television sets and automobiles they use. If the television set or the automobile lasts a long time, it is said to be reliable.

Up to now we defined quality as "conformance to specifications" at one point in time. This is a static definition of quality. It says nothing of how a product functions under varying environmental conditions over time. Excessive variations in materials, operators, methods, machines, or environment can cause premature failure.

The difference between quality as a static concept and reliability as a dynamic concept can be illustrated by an example. An automobile shock absorber is mainly a large spring. A new automobile shock absorber must compress and be able to sustain a certain load at one point in time. Over time, the shock absorber is compressed many times. The quality question of a new absorber is, how far does it compress? The reliability question is, how many compressions can the absorber withstand before it fails and the spring breaks?

Reliability is the probability that a product will perform its intended function without failure under desired operating conditions over a specified period of time. This definition has three elements that have important implications which we discuss briefly: intended function, desired operating conditions, and stated period of time.

Intended Function. **Intended function** is the ability of a product to perform a task or operate upon command. It often depends on multiple perspectives, including the manufacturer's, the regulator's, the customer's, and the user's perspective. A person reading a book wants light when the switch is flipped on. A driver wants an automobile to run indefinitely. The automobile's manufacturer wants it to run without repair during its warranty period. The designer wants a product to function according to specifications. The government wants a product to be safe. Consumer groups want it to be safe, reliable, and fairly priced. The design engineer is therefore responsible for obtaining sufficient information from marketing to design a product that satisfies both present and future customer requirements.

Desired Operating Conditions. Product reliability also depends on the environment in which the product operates. A benign environment can extend product life while an adverse environment can cause premature failure. An environment in which a product operates can be inside a structure or outside, exposed to the elements. The external environment exposes a product to heat, water, salt, wind, humidity, and vibration. Internal conditions, including factors such as heat, vibration, dust, and power surges, can also impair operations. Electrical components are especially susceptible to these hazards. Power surges can easily disrupt electrical equipment. Sensitive electrical equipment has a surge protector to protect it against power spikes.

Any heat, regardless of whether internal or external, can also freeze mechanical components and shorten electrical component life. Heat is dissipated in mechanical and electrical components by having water or oil pass over heated surfaces.

Stated Period of Time. The last element in the definition of reliability is that the product must operate as intended over a stated period of time. This may be as short as several hours or as long as thousands of hours.

The stated period of time is specified by the design engineer and dictated by the needs of the customer. A product designer or industrial customer may specify the reliability of a typical light bulb, such as a 60-watt bulb, to be 99.9% for 4000 hours of operation under ambient conditions. Reliability testing of a number of light bulbs has indicated that the typical bulb is 99.9% reliable for 6000 hours of operation inside a warm environment. The company also knows that the average light bulb is used 2000 hours per year. Thus the manufacturer feels secure that it can issue a warranty for 4000 hours of operation or 2 years, whichever comes first.

RELIABILITY MEASUREMENT

Reliability is expressed as a probability from 0.0 to 1.0, where 0.0 means that the product will fail and 1.0 that the product will operate as intended over a stated time period.

Reliability is expressed in the following terms:

- **Failure rate** is the mean, or average, number of failures in a given interval of time or, in other words, the probability that a product will fail within a specified period of time.
- **Mean time between failures (MTBF)** is the mean, or average, time between failures of products that are repaired and maintained. Failure rate and mean time between failures are reciprocals of each other.
- **Mean time to failure (MTTF)** is the mean, or average, time to failure of products that are not repaired or maintained. MTTF can be expressed in terms of the number of hours of use (such as 1000 hours) or in terms of the percentage failed during a period of time (10% failed after 1000 hours of use).

Repairable or Nonrepairable

The preceding definitions differentiate between a product that is repairable and one that is not repairable. Mean time to failure is used for a product that has not been repaired and, when it does fail, is not replaced by a good product. Mean time between failures is used to calculate the reliability of a product that can be repaired. For a repairable product, failure is assumed to occur at a constant rate.

Light bulbs, shock absorbers, and space satellites are examples of nonrepairable products. Only one failure can occur with such a product, then it is scrapped. A repairable product can be maintained and fixed. Examples of repairable products include computers, compressors, and automobile transmissions.

Sometimes a product is considered both repairable and nonrepairable. For example, a rocket engine on a test stand is considered repairable. It becomes nonrepairable when it is used to launch a satellite. Thus it has become customary to use mean time between failures for both repairable and nonrepairable products.

EXAMPLE   TALCO manufactures microwave ovens. In order to develop warranty guidelines, TALCO randomly tested 10 microwave ovens continuously to failure. The failure information of the 10 ovens is as follows:

Microwave	Hours	Microwave	Hours
1	2300	6	1890
2	2150	7	2450
3	2800	8	2630
4	1890	9	2100
5	2790	10	2120

What is the mean time to failure of the microwave ovens?

Solution
The total number of hours the microwave ovens were used is 23,120. On the average, a microwave oven failed every 2,312 hours. The failure rate and the mean time to failure (MTTF) are calculated as follows:

$$\text{MTTF} = \frac{23,120 \text{ hours}}{10 \text{ microwave ovens}} = 2312 \text{ hours}$$

The mean product life of the microwave ovens is defined in terms of their mean time to failure because no maintenance is performed on the ovens.

Types of Failure

Failure is the ability of a product to perform its intended function. Product failure is usually categorized by type, degree, and time. For example, a **catastrophic failure** is sudden and complete. A **degradation failure** is gradual and partial. What do these terms mean? A sudden failure occurs without any warning. A **gradual failure** occurs over time and may be anticipated by prior history. A **partial failure** occurs because of a minor or major deviation or nonconformance. It causes a malfunction but does not cause a complete product failure. A critical defect causes a complete failure.

Sources of Reliability Information

Reliability information is obtained from many sources. The two most common ones, which are discussed here, are customer feedback and product testing.

Customer Feedback. Information obtained from customer surveys or through failed products provides valuable usage information. Market research and customer surveys provide subjective information of customer perceptions. Field failures indicate how, why, and when the product failed. If the product was inadequate for its intended function, it will be redesigned. If the product is misused or abused, a warning label or operating and maintenance instructions will be placed on the product.

Product Testing. Tangible products do fail over time. The nature and timeliness of the failure is important. If a product fails and someone is hurt, the injury can be very costly if it goes to trial. Also, if a product fails during its warranty period, then it is very costly to repair or replace. If a failure occurs after the warranty period has elapsed, the manufacturer is usually not liable for any direct costs, but the indirect costs of loss of reputation and lost sales can be equally damaging.

Products, assemblies, subassemblies, and components are reliability tested to determine their limits. Product testing allows results to be monitored, measured, and recorded under controlled conditions. Tests can simulate conditions under which a product operates as well as certain types of failure. Common reliability tests include vibration, temperature cycling, and overstressing.

Test results are reliable, but testing is expensive and time consuming. However, if tests are conducted properly, a product can be designed to be more reliable.

BATHTUB CURVE

Reliability over Time

A laptop computer has just been developed. The designers want to know its reliability, so a simple test is developed. One hundred computers representative of the entire production lot are tested at the same time. The test is conducted in a lab that approximates customer usage. Failures are tracked and noted. As failures occur, the computers are not replaced, so that the number of functioning computers diminishes steadily. Certain assumptions are made to facilitate the calculation and subsequent explanation.

After about 20 hours of continuous use, 12 computers have failed to function. Most of the failures occurred between 15 and 20 hours. The test engineers calculate the failure rate during this period as follows:

$$\text{Reliability over 20 hours} = \frac{\text{number functioning}}{\text{total number}}$$

$$= \frac{88}{100} = 0.88$$

The test engineers notice that the failure rate decreases after the 20th hour and becomes constant. Failures occur sporadically for the next 2000 hours of use. During this time, only five failures occur. There does not seem to be a likely cause for the failures. Failures are random and intermittent.

From 2020 to 2200 hours, more machines start failing in clusters. By 2200 hours, 45 machines have failed. The engineers believe the failures are no longer random but are caused by a critical electronic component. This failure pattern, as well as many others, follows a three-phase pattern, which in graphic form is shaped like a bathtub.

What Is a Bathtub Curve?

The **bathtub curve** is illustrated in Figure 9.1. It shows the pattern of failure for many products, especially electronic components. As the example indicates, the failure pattern of many products can be graphed by testing the products to failure. The bathtub curve is divided into three parts: infant mortality, useful life, and wearout.

Infant Mortality. **Infant mortality** is also called early failure, burn-in, break-in, or debugging stage. In this stage, the failure rate is high but decreases gradually. During this period, failures occur because engineering did not test products sufficiently, or manufacturing made defective products. Failure is due to nonconforming components, where the probability of failure depends on the length of time the part was operating.

Failure may also occur because the product was overly used, misused, or abused, or because it was operated in an environment in which it was not intended to function. In these circumstances, either the manufacturer or the user may be at fault. The manufacturer may not have designed the product properly, may have limited its use to certain conditions, or may not have informed the user of proper maintenance or operation. The user may be at fault by having misused or abused the product or by simply not having read the attached labels and instructions.

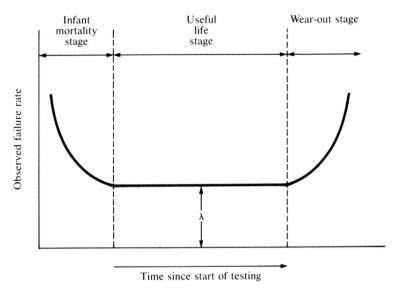

FIGURE 9.1
Product failure pattern illustrated by bathtub curve.

Useful Life. Useful life is the middle stage of the bathtub curve. It is characterized by a flat, constant failure rate. Design flaws have been eliminated, manufacturing operations are under statistical control, and defect-free products are being produced. The product has been redesigned to enhance reliability, maintainability, and performance.

Product reliability with a constant failure rate can be predicted by the following formula:

$$R = e^{-\lambda t}$$

where R = reliability
e = natural logarithm
t = time, hours
λ = failure rate

Components with very high MTTFs are used in order to increase the reliability of the total product. However, even these components will fail randomly during this stage.

Wearout. Wearout is the final stage where the failure rates increase as the products begin to wear out because of age or lack of preventive maintenance. Components may fail because of wear due to cyclic loading, overstressing, or simply aging.

Extending the Operating Life

Products, especially electronic components, can have their useful life extended by proper design, conscientious assembly, and careful handling (Figure 9.2). Theoretically it is possible to extend the useful life of a product indefinitely through preventive maintenance.

Individual components may conform to specifications, but when they are added to other parts in an assembly, they may interact in unanticipated ways,

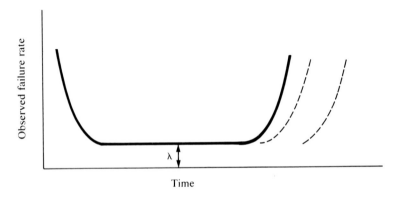

FIGURE 9.2
Extended bathtub curve.

thereby causing premature failures. Electronic components must be assembled carefully as they are stuffed into a circuit board. Solder must provide sufficient contact and heat transfer. Finally components should be tested carefully, or a static discharge will short-circuit any reliable design.

Finally, electronic components should be burn-in tested, that is, tested under conditions that lead to failure. As parts fail, they are identified and replaced. The reliability of the final product is thus improved. If a certain component shows a history of failure, then it may have to be modified, replaced, or redesigned.

Modeling Reliability

The reliability of a component or product can sometimes be approximated by equations. Engineers try to determine the failure patterns of the three stages of the bathtub curve by performing laboratory tests or through obtaining customer use information. Laboratory tests can show whether the failure pattern is shaped like the bathtub curve. If it is, the slope of the early failure and wearout curves can be calculated as well as the length of the useful period.

However, caution must be exercised in modeling reliability. Laboratory testing may not approximate real-world usage. Customer use information and examination of failed products do not reveal how the product was used, by whom, or for how long. In the real world, many variables act together to decrease a product's reliability, and these cannot be expressed by an equation. A laboratory experiment carefully controls the variables that may limit a product's reliability.

EXAMPLE   An industrial machine compresses natural gas into an interstate gas pipeline. The compressor is on line 24 hours a day. If the machine is down, a gas field has to be shut down until the natural gas can be compressed. Downtime is very expensive.

The company is shopping for a new compressor and wants to know the operational reliability for 2000 hours of continuous service. The vendor knows the compressor has a constant failure rate of 0.000001 failures per hour.

Solution
The key to solving this problem is realizing that the compressor has a constant failure rate. This indicates which equation is to be used.

Given $t = 2000$ hours and $\lambda = 0.000001$ failure per hour, the reliability is

$$R = e^{-\lambda t}$$
$$= e^{-0.000001(2000)} = 0.998$$

for 2000 hours of operation.

EXAMPLE   If the operational reliability for a mainframe computer is 0.999 for 1000 hours of continuous operation, what is its constant failure rate?

Solution

The constant failure rate equation must be manipulated to solve this problem:

$$R = e^{-\lambda t}$$

Given R = 0.999 and *t* = 1000 hours, determine λ.

$$\ln R = -\lambda t$$

$$-\frac{\ln R}{t} = \lambda$$

$$\lambda = -\frac{\ln 0.999}{1000} = 0.000001 \text{ failure per hour}$$

DESIGNING FOR RELIABILITY

Reliability Factors

Earlier in this chapter we mentioned that a product designer may specify that a 60-watt light bulb be 99.9% reliable over 4000 hours of operation in ambient conditions. By specifying light bulb reliability, the engineer is defining its design parameters.

The design engineer has enough information to design a cost-effective light bulb that will last 6000 hours. He or she knows from laboratory testing that few bulbs will fail within a warranty period of 4000 hours or 2 years of operation under specified operating conditions.

Enhancing design reliability is a matter of balancing many factors. The design engineer must identify the most important factors, optimize these, and minimize any constraints. For example, an optimum product would try to satisfy everyone by having the following characteristics:

☐ High reliability under extreme conditions
☐ Easily maintainable under extreme conditions
☐ Inexpensive
☐ Safe under any condition
☐ Easily used by anyone
☐ Lightweight

It is impossible to develop a product that satisfies all of these factors. There are tradeoffs. For example, reliability may be increased by adding parts, which however will add cost and weight to the final product.

Reliability Design

The design engineer must plan for future events and contingencies in both industrial and consumer products. Reliability is usually specified more precisely in industrial, commercial, military, and aerospace products. These products are used in extreme environmental conditions and must operate upon command.

Consumer products are generally not as reliable as industrial products because they are not used in such harsh environments. However, consumer products, such as automobiles and lawn mowers, must still be safe.

Murphy's law, "if it can fail, it will fail at the worst possible time," applies to the study of reliability. New products, consisting of many components, are highly complex. The probability of failure is high because of abuse or random disturbance. The probability of failure increases as users want products to function in unintended ways in extreme environmental conditions. Products are pushed beyond performance limits and are operated by people who may not have read directions.

Simple, universal rules have been developed for designing reliable products*:

- ☐ Use as few components as possible.
- ☐ Use reliable components.
- ☐ Use standard components.
- ☐ Configure components into logical systems.
- ☐ Derate components and system.
- ☐ Institute preventive maintenance and repair.

Use as Few Components as Possible. Modern products consist of many components that interact in complex ways. The possibility of failure increases as the number of components increases. Often such an increase in components and complexity cannot be avoided because the user wants the product to perform additional and complicated functions. One solution to reducing product complexity is to eliminate unnecessary components.

Use Reliable Components. A product is only as good as its weakest component. A complex product may consist of thousands or millions of components. A small part, such as a rivet on an airplane, may cause a bulkhead to fail.

A product's reliability can be increased if each component's reliability is known. For example, if the engineer knows that a typical component or assembly will fail within 2000 hours, the engineer can specify that the component be replaced or the assembly maintained every 1800 hours, thereby forestalling a product failure.

Use Standard Components. New products tend to have performance, fit, or reliability problems. Reliability can be increased by choosing standard, commercial parts of known reliability and staying away from specially designed components.

Standard components can be ordered from manufacturers' catalogs and are known to comply with industry, national, or federal standards. Data sheets and commercial specifications of standard components specify failure rates and per-

---

* R. Caplen, *A Practical Approach to Reliability* (Business Books, London, 1972), pp. 70–86.

formance curves for time in use. These components are cheap and have been tested thoroughly. Any initial flaws have been designed out of these components.

Configure Components into Logical Systems. A final product can be configured in many ways. Depending on their reliability, individual components can be placed in series, parallel, or series/parallel. As a rule of thumb, reliable components are placed in series and less reliable components are placed in parallel. The reason for this is discussed in the next section.

Derate Components and System. A product may be derated, that is, the manufacturer wants it to operate at a lower performance level. Derating is similar to including a **safety margin.** The manufacturer has tested prototypes until failure to determine the upper performance, strength, or durability limits. The upper limit is then reduced to an acceptable safety level, which becomes the recommended operating level.

Derating ensures safety by allowing a person to use a product slightly above recommended operating limits without having it fail catastrophically. Also, derating is useful for determining the warranty period. If a manufacturer knows through testing at a higher performance level that 99.9% of the products will last through 1000 hours of operation, the warranty will be specified at 1000 hours. If the distribution of failure of a number of products is shaped like a normal curve, the normal curve can be used to estimate the percentage of the population that is going to fail.

Institute Preventive Maintenance and Repair. Preventive maintenance is the same as scheduled or planned maintenance. It tries to anticipate when a product will fail by calculating component failure rates and replacing or servicing parts just before they fail. When a product fails, the cost of repair is more than the cost of routine maintenance. Often when a product fails, components are destroyed and must be replaced entirely. Preventive maintenance may only entail an adjustment, check, lubrication, or exchange of a minor component.

COMPONENT CONFIGURATION

A complex product may consist of multiple levels of assemblies, subassemblies, systems, subsystems, and components. Each product level can be configured in different shapes. Reliability then becomes a function of the configuration and individual reliability of the components. Reliability engineers try to enhance product reliability by selecting the proper components and configuring them in reliable and cost-effective ways.

Components, systems, or assemblies can be placed in three basic configurations: in series, in parallel, or in series/parallel.

Series Configuration. Components in series are placed one after another. If one component should fail, the whole system will fail. Like links in a chain, if one link breaks, the entire chain fails.

EXAMPLE The following components are placed in series: $R_1 = 0.90$ and $R_2 = 0.90$. What is their equivalent, or total, reliability?

Solution

$$R_t = R_1 \times R_2$$
$$= 0.90 \times 90 = 0.81$$

This product can be drawn schematically as illustrated in Figure 9.3. Each box represents a component with a known reliability. As can be seen, total reliability is less than the reliability of each component.

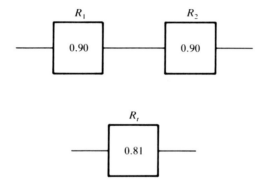

FIGURE 9.3
Series configuration.

Series reliability can be expressed as follows:

$$R_t = R_1 \times R_2 \ \cdots \ \times R_n$$

where R_t = total reliability
R_1 = reliability of first series component
R_2 = reliability of second series component
R_n = reliability of nth component

When several components are placed in series, the total system reliability decreases quickly. Assuming that each component has the same reliability of 0.90, the total reliability of up to 10 components in series is given in the following table:

Number of Components	Equivalent Reliability
0	0
1	0.900
2	0.810
3	0.729
4	0.656
5	0.590

Number of Components	Equivalent Reliability
6	0.531
7	0.478
8	0.430
9	0.387
10	0.349

Parallel Configuration. To increase reliability, components are placed in parallel. This provides redundancy so that if one component fails, a parallel component will operate and keep the system functioning.

In a two-component circuit, either component can operate, or both components can operate as intended. This is similar to the addition rule for nonmutually exclusive events. As you may recall, the addition rule is expressed in terms of the following probabilities:

$$P(\text{component 1 or component 2}) = P(1) + P(2) - P(1)P(2)$$

The equation for a two-component circuit is

$$R_t = R_1 + R_2 - R_1 R_2$$

The general formula for calculating the total reliability of a parallel circuit is

$$R_t = 1 - (1 - R_1)(1 - R_2) \cdots (1 - R_n)$$

where R_t = total reliability
R_1 = reliability of first leg
R_2 = reliability of second leg
R_n = reliability of nth leg

EXAMPLE Parallel components have the following reliabilities: $R_1 = 0.90$ and $R_2 = 0.90$. Find the total reliability.

Solution
As Figure 9.4 illustrates, the reliability of parallel components is higher than the reliability of any single component or similar components in series.

$$R_t = 1 - (1 - R_1)(1 - R_2)$$
$$R_t = 1 - (0.10)(0.10)$$
$$= 0.99$$

Series/Parallel Configuration. Electric circuits or mechanical assemblies are often designed in a series/parallel configuration. Low-reliability components are placed in parallel and components with relatively high reliability are in series.

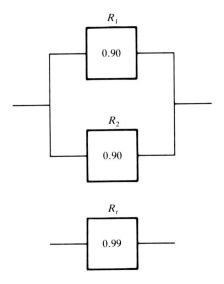

FIGURE 9.4
Parallel configuration.

Series/parallel configurations look complex on paper, but are relatively easy to solve. The objective is to simplify the configuration and calculate the total reliability. The process follows these steps:

☐ Step 1. Calculate the equivalent reliability of components in series within any parallel configuration.
☐ Step 2. Reduce each parallel configuration to an equivalent series reliability.
☐ Step 3. Multiply each series reliability until the smallest total reliability is derived.

EXAMPLE The components illustrated in Figure 9.5 have the following reliabilities: $R_1 = 0.90$, $R_2 = 0.90$, $R_3 = 0.90$, $R_4 = 0.90$, and $R_5 = 90$. The first two components are in series, the last two in parallel. Find the total reliability.

Solution
This is a simple series/parallel circuit.

Step 1: The upper leg has two components in series, R_3 and R_4. The equivalent reliability R_{eq1} is

$$R_{eq1} = R_3 \times R_4$$
$$= 0.90 \times 0.90 = 0.81$$

Step 2: The next step is to find the equivalent reliability of the parallel circuit, R_{eq2}:

$$R_{eq2} = 1 - (1 - R_5)(1 - R_{eq1})$$
$$= 1 - (1 - 0.90)(1 - 0.81) = 0.98$$

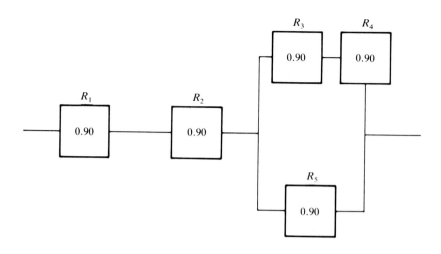

Step 1:

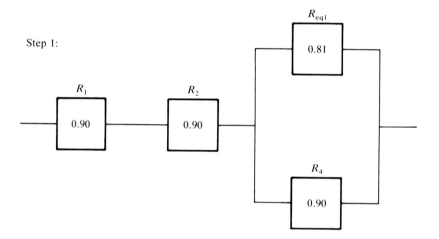

Step 2:

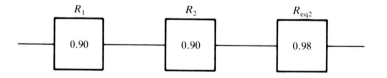

Step 3:

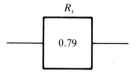

FIGURE 9.5
Series/parallel configuration.

Step 3: In the final step we multiply the individual series reliabilities to find the total reliability R_t:

$$R_t = R_1 \times R_2 \times R_{eq2}$$
$$= 0.90 \times 0.90 \times 0.98 = 0.79$$

The method used to solve this circuit can also be applied to more complicated circuits.

MAINTAINABILITY, REPAIRABILITY, AND AVAILABILITY

What Is Maintainability?

Maintainability is the ease by which a product can be serviced or maintained. Operational delays represent expense. If a product is maintained and repaired periodically, it is theoretically possible to extend its useful life indefinitely. So an important element in designing a reliable product is to design it for maintainability.

Preventive maintenance (PM), like any prevention program, must be planned. Reliability at each stage of the bathtub curve can be calculated to predict failure before it occurs. PM is expensive. Machinery is down and production stops. Depending on the nature of the operation, PM is always scheduled when it does not disrupt normal operations, such as during the graveyard shift or during holidays.

What Is Repairability?

Repairability is the ease by which a product can be repaired. Repair and maintenance are different concepts. A product is maintained while it is operational so that premature failure may be prevented. A product is repaired when it fails.

When a product fails, the problem should be located quickly and repaired easily. Sophisticated equipment may have internal diagnostic procedures such as a software program that can check the function of the product. Mechanical components indicate problems through instrumentation, lights, or even noise. Some automobiles have brakes that chirp when the pads have reached a dangerous level of wear.

Fault location must be quick and effective, not haphazard. A technician cannot check every element of a product. Convenient test points and self-diagnostics speed up the locating of faults.

Once the cause of a fault is found, it must be corrected. Products are designed so that panels, assemblies, or modules can be pulled out and replaced easily. Parts with a history of high failure are identified and designed out of the product. The product should also be easy to disassemble and assemble.

Many elements of an automobile cannot be repaired because they are in sealed units that cannot be disassembled without breaking the unit and voiding any warranties. In this manner automobile manufacturers want to limit any liability exposure of a poor repair job by an incompetent mechanic.

What Is Availability?

Availability is the amount of time a piece of equipment is free to be used. Availability introduces a new concept of time to repair. Mean time to repair (MTTR) is the mean, or average, time it takes to repair a product after it has failed. Critical products usually have a low mean time to repair. These products can be diagnosed and repaired quickly by focusing on modular design and ease of maintenance.

Availability is an important characteristic of tactical military aircraft, which must be operational much of the time. In banking, large mainframe computers must be available to process large numbers of checks.

Availability is directly influenced by the repairability of the product. If a product is difficult to repair, the mean time to repair increases:

$$\text{MTTR} = \frac{\text{total number of hours spent on repairs}}{\text{total number of repair jobs}}$$

Availability can now be expressed by the following formula:

$$\text{Availability} = \frac{\text{MTBF}}{\text{MTBF} + \text{MTTR}}$$

where MTBF = mean time between failures
      MTTR = mean time to repair

HUMAN ENGINEERING

Ease of repair and maintenance are important considerations in the design of a product. Both elements are associated with the study of ergonomics.

Ergonomics is the engineering study of fitting a product to human abilities rather than vice versa. For example, simple products, such as orange juice containers, have a handle and are shaped to be held and poured easily.

Ergonomics considers many factors. The following are discussed in this text: cultural, packaging, working conditions, layout of control and equipment, and instrumentation.

Cultural. The same product may be shipped, handled or stored differently depending on the environment and needs of the end user. Products used in the Orient must consider humidity, heat, religion, customs, and operating directions.

Packaging. Proper packaging can make a product safe and easily recognizable, and can maintain its quality. Soap and shampoos are dispensed from easily held plastic bottles to reduce slipping and sliding in the shower. Packaging creates brand awareness. When consumers are confronted by competing brands, packaging differentiates a product from its competition. Finally, packaging keeps product contents fresh during shipping, storage, and display.

Working Conditions. Adverse working conditions or a difficult piece of machinery can cause operator fatigue and subsequent accidents. In a cold environment the operator will focus on discomfort rather than on the task at hand. The attention span decreases and risks increase. Difficult working conditions may also involve a cold, dirty, or constricted work space; or difficulty may be caused by constant bending, lack of proper tools, or poorly organized equipment. For example, farm equipment manufacturers are aware of the importance of a safe and warm environment for the farm equipment user. Often the equipment user is also the buyer in a family-owned farm. The piece of equipment may be used by a migratory worker who cannot read English instructions or by a youngster who does not want to read instructions. Tractors are therefore built simply and ruggedly such that they are easy to use, repair, and maintain.

Layout of Control and Equipment. Equipment controls are laid out in an orderly and logical manner. For example, if the electrical starter on a lawn mower is near the blade, there is a high probability of a person being hurt. Controls in a car are placed so that most people can reach them comfortably while driving. Most dials and indicators are positioned so they can be read easily. Controls placed approximately at eye level result in less strain and parallax error.

Instrumentation. The manner in which instrumentation is designed affects its understandability. For example, dials on meters can be digital or analog. Digital watches have a numerical readout; analog watches have hands pointing to numbers. Digital watches were popular 10 years ago because they were a novelty and cheap. But analog watches had a resurgence of interest because they are quicker and easier to read. People know what it means when the big and small hands are in certain positions. Even in sophisticated electronic instrumentation dial-type meters are preferred over digital ones when quick interpretation is needed.

SUMMARY

☐ Reliability is long-term quality. Technically, reliability is the probability that a product will perform its intended function under desired operating conditions over a specified period of time.

☐ Reliability is measured in terms of failure rate, mean time between failures (MTBF), and mean time to failure (MTTF).

☐ The bathtub curve illustrates the failure pattern of many products. This curve is divided into three elements: infant mortality, useful life, and wearout.

☐ Designing to enhance reliability entails balancing many tradeoffs, including price, weight, safety, ease, and reliability. There is no optimum product that will incorporate all of these requirements. Given specific design parameters, the engineer can enhance reliability by using fewer, reliable, standard components that are configured in a manner that improves a product's life.

☐ Components can be configured in a series, parallel, or series/parallel manner. The actual configuration depends on the reliability of the system, given certain design parameters.

☐ Other important "abilities" are maintainability, repairability, and availability. Maintainability is the ease by which a product can be maintained. Repairability is the ease by which a product can be serviced. Availability is the amount of time a piece of equipment is operational.

☐ Ergonomics is the study of fitting a product to human abilities. Ergonomics is a function of many factors, including cultural, packaging, work conditions, layout of controls, and instrumentation.

KEY TERMS

availability   Amount of time a piece of equipment is free to be used.

bathtub curve   Graph showing failure patterns for many products.

catastrophic failure   Sudden and complete failure.

degradation failure   Gradual and partial failure.

ergonomics   Engineering study of fitting a product to human abilities rather than vice versa.

failure rate   Mean, or average, number of failures in a given interval of time.

gradual failure   Failure occurring over time; may be anticipated by prior history.

infant mortality   First part of the bathtub curve; failure rate is high but decreases gradually; also called burn-in, break-in, or debugging phase.

intended function   Ability of a product to perform a task or operation upon command.

maintainability   Ease by which a product can be serviced or maintained.

mean time between failures (MTBF)   Mean, or average, time between failures of products that are repaired and maintained.

mean time to failure (MTTF)   Mean, or average,

time to failure of products that are not repaired or maintained.

parallel components   Components placed side by side.

partial failure   Failure causing a malfunction but not a complete failure.

preventive maintenance   Maintenance of a machine or product to prevent premature failure.

reliability   Long-term quality; the probability a product will perform its intended function without failure under desired operating conditions over a specified period of time.

repairability   Ease by which a product can be repaired.

safety margin   Measure of safety in case the product is operated beyond specified limits or under conditions that were not intended; similar to derating.

series components   Components placed one after another.

useful life   Middle part of the bathtub curve; characterized by a flat, constant failure rate.

wearout   Last part of the bathtub curve; product failure rate as products begin to fail because of age.

QUESTIONS

1. Why is reliability becoming more important? Give examples where it is imperative.

2. What is reliability? Discuss the important elements of the definition.

3. Adverse environmental conditions can affect product reliability. Discuss how reliability can be enhanced.

4. What is the difference between mean time between failures and mean time to failure?

5. What is the difference between the following types of failure: catastrophic, degradation, sudden, and gradual?

6. What is a better source of reliability information, customer feedback or product testing? Discuss the merits of each.

7. Is the bathtub curve a reliable model for forecasting product failures?

8. How can the useful life of a product be extended?

9. It is not possible to develop an optimum product with all the characteristics that people want. Discuss this concept and give examples.

10. Explain three methods for enhancing product reliability.

11. A principle often heard in reliability is the KISS principle. It stands for Keep It Sweet and Simple. Discuss this concept of reliability engineering.

12. With proper preventive maintenance, is it possible for a product to have an infinite life?

13. Why are parallel circuits used in certain types of design? Give an example where a parallel circuit may be used.

14. Define repairability and maintainability. What is the major difference between the two?

15. Why is availability such an important factor in the military?

16. What is ergonomics? Give an example of an ergonomic design in your car.

PROBLEMS

1. If a component has a failure rate of 1.2×10^{-5} failures per hour, what is its reliability for 100 hours? for 1000 hours? If there are 100,000 components in the test, how many failures are expected in 1000 hours?

2. A component has a failure rate of 0.005 failures per hour and a reliability of 0.99. What is the mean time between failures?

3. If a component has a failure rate of 2.5×10^{-6} failures per hour, what is its reliability for 200 hours? for 3000 hours? If there are 10,000 components in the test, how many failures are expected in 3000 hours?

4. What is the reliability of a product with a failure rate of 1.7×10^{-5} for operating periods of 1000, 10,000, and 100,000 hours?

5. A component has a mortality of $0.000015e^t$. Calculate the failure rate and the MTBF for 2000 hours.

6. A component has an MTBF of 10^5 hours. Calculate the reliability for 100, 1000, and 2000 hours.

7. A component has an MTBF of 1.5×10^6 hours. Calculate the reliability for 1000, 10,000, and 100,000 hours.

8. Five components in series have individual reliabilities of 0.99, 0.99, 0.90, 0.90, and 0.85. What is the total reliability of the system? Draw a schematic diagram of the system.

9. Four components in series have individual reliabilities of 0.99, 0.99, 0.99, and 0.80. What is the total reliability of the system? Draw a schematic diagram of the system.

10. Three components in parallel have individual reliabilities of 0.99, 0.99, and 0.99. What is the total reliability of the system? Draw a schematic diagram of the system.

11. Three components in parallel have individual reliabilities of 0.99, 0.90, and 0.85. What is the total reliability of the system? Draw a schematic diagram of the system.

12. Two components in parallel have individual reliabilities of 0.99 and 0.95. This set of parallel components is in series with another set of parallel components with individual reliabilities of 0.85 and 0.80. What is the total reliability of the system? Draw a schematic diagram of the system.

13. Two components in parallel have individual reliabilities of 0.80 and 0.80. This set of parallel components is in series with another set of parallel components with individual reliabilities of 0.90 and 0.90. What is the total reliability of the system? Draw a schematic diagram of the system.

10

Quality Economics

In this chapter we discuss the theoretical basis of quality costing and present practical methods for analyzing these costs. Specifically, this chapter covers quality costing, maintenance costs, reliability costs, Taguchi costing, and investment analysis.

Investment decision making is not a traditional topic in quality control or assurance texts. It is thought to be an area of business administration and management. However, many decisions in quality engineering and management require this type of analysis to justify a new piece of equipment or approve a quality improvement project.

Importance of Reliable Cost Information

A company-wide quality management (CWQM) program succeeds or fails based on the accuracy and immediacy of the cost information, which ensures reliable decision making. Quality decision making involves the deployment of scarce individual, corporate, and societal resources in an atmosphere where there are seemingly infinite wants. Sometimes people believe quality should be pursued at any cost. But nowadays most quality improvement projects must compete against other projects, and scarce resources, whether time, material, people, or finances, must be balanced against other competing uses and demands. However, there are instances where quality should be pursued regardless of cost, namely, where safety or health issues are involved. Quality appraisal and prevention should be pursued in engineering, constructing, operating, and maintaining a nuclear power plant, for example. The risk of not doing so is catastrophic failure.

Quality cost information is a management tool to monitor and control quality costs during production, evaluate competing projects, and ensure on-budget and on-time completion. Specifically, quality cost information can be used to:

☐ Determine the hidden costs of customer dissatisfaction
☐ Determine the cost of a field failure
☐ Determine the costs of producing a product
☐ Establish a benchmark or baseline level of quality
☐ Make accurate decisions
☐ Compare achieved objectives against stated objectives
☐ Compare returns of one quality improvement project against those of another
☐ Measure the effectiveness of any corrective action or quality improvement
☐ Evaluate management performance
☐ Allocate scarce resources to projects promising the greatest return

CONFORMANCE AND NONCONFORMANCE COSTS

Quality costs are divided into two broad categories, conformance and nonconformance costs. **Conformance costs** are incurred in attempting to conform to product specifications or trying to satisfy customer requirements. Conformance-related costs are further broken down into prevention and appraisal costs. **Prevention costs** are costs associated with planning, managing, training, and measuring quality. **Appraisal costs** are costs associated with evaluating, inspecting, and testing products.

Nonconformance costs are incurred by not being able to comply with product specifications or satisfy customer requirements. These costs result directly from a product flaw or defect or from not complying with a procedure. They are further broken down into internal failure and external failure costs.

Internal failure costs include the cost of rework, scrap, machine downtime, and product testing, while **external failure costs** result from poor procured material, warranty service, and product liability.

Quality Cost Curves

The relationship between conformance and nonconformance costs can be shown by **quality cost** curves. These curves are theoretical and difficult to apply to specific quality situations. However, for explanatory purposes, these curves illustrate the relationship between conformance and nonconformance costs.

The quality cost diagram in Figure 10.1 shows three curves, conformance (prevention and appraisal), nonconformance (internal and external failure), and total cost. The horizontal axis indicates the distribution of nonconformance and conformance costs. If products and services are 100% conforming, prevention and

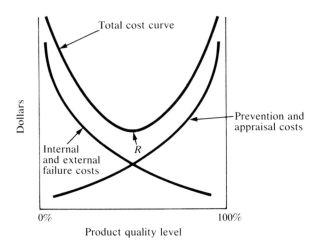

FIGURE 10.1
Quality cost curves.

appraisal costs are very high and internal and external failure costs approach zero. Similarly, if products and services are 100% nonconforming, prevention and appraisal costs approach zero and internal and external costs are theoretically very high.

There is an inverse relationship between conformance and nonconformance costs. Internal and external failure costs are incurred because products fail and services dissatisfy customers. Balanced against these costs are prevention and appraisal costs, which are expended to reduce internal and external failure costs. If systems for prevention and appraisal did not exist, nonconformance costs would increase. To reduce these costs, resources are dedicated to prevention and appraisal.

Minimum Point. The theoretical minimum total cost, point R, is the lowest point of the total cost curve. This point is the balance between nonconforming and conforming costs.

The shape of the total cost curve depends on the requirements of the particular industry or product. For example, in the nuclear power industry, prevention and maintenance are emphasized to reduce the possibility of a power outage or system failure. A major failure, such as a nuclear meltdown and emission of radioactivity into the atmosphere, is to be averted at any cost. To ensure a low probability of catastrophic failure, prevention and appraisal costs increase tremendously.

Reliablity Cost

Reliability and maintainability costs can be analyzed in a similar way as quality costs. Reliability and maintainability are comparable to conformance and nonconformance. This relationship is illustrated in Figure 10.2.

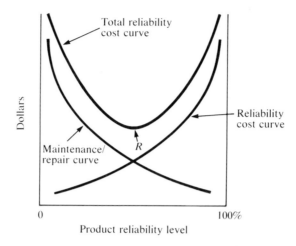

FIGURE 10.2
Cost of reliability.

The cost of reliability has an inverse relationship to the cost of maintenance and repairs. On the left, reliability cost is low while maintenance and repair costs are high. When product reliability is low, maintenance costs due to breakdown of machinery are high. On the right, the costs of achieving reliability are high while maintenance and repair costs are low.

If reliability and maintenance costs are summed, the result is the total reliability cost curve. The minimum level of cost is at the low point of the total cost curve, point R. The optimum level of cost is probably somewhere to the right of this point because of the higher cost of achieving reliability. Increased reliability is often more important than slightly increased costs because of the importance of safety and the litigious nature of our society.*

COST MEASUREMENT SYSTEM

Implementing a cost measurement system is a systematic process that proceeds along the following steps: obtain management support, identify improvement project, identify baseline costs, implement quality measurement system, and interpret results.

Obtain Management Support. Before a quality cost measurement system is initiated, an organization must be aware of the importance of quality costs and the need for such a system. The commitment of the organization and its resources is required if the system is to be developed and implemented. Developing a quality cost measurement system without the commitment and active participation of management leads to sure failure.

* R. Caplen, 1972, *op cit.*

A quality cost measurement system opens windows into different parts of the organization, thereby identifying opportunities for cost reductions. If top management supports the quality measurement system, the effort has a good chance of being accepted and implemented by middle and lower level management.

A quality measurement effort can be threatening. Department managers often are protective of their operations and how money is allocated. Unfortunately a protective department manager gives the impression he or she is trying to conceal something.

Furthermore, a quality measurement system has a greater chance of being accepted if it is supported by the accounting organization. This department already understands the existing chart of accounts and has a political reason for backing a quality cost system.

Identify Improvement Project. The quality cost measurement system should be coupled with a quality improvement project. Such a project may involve training personnel, buying a new piece of machinery, establishing SPC in a process, or designing a product. In this way costs, benefits, and improvements can be tracked. It is hoped that the improvement project would demonstrate that quality cost information can benefit the organization by identifying areas where it can reduce costs or improve productivity. If the improvement project is successful, then a wider quality cost measurement effort can be justified.

Identify Baseline Costs. Once an improvement project is instituted in an isolated area, existing quality-related costs should be identified. These costs are baselines, or benchmarks, from which to measure any improvement. The costs associated with attaining a desired level of quality are then compared against accrued benefits. Any incremental improvement is used to justify a proposed project or to approve future improvement projects. Top management will realize the usefulness of these costs to track departmental, business unit, or corporate objectives. Without measurements against baselines, corporate quality objectives cannot be met and a satisfactory level of performance cannot be achieved.

Implement Quality Measurement System. Once an improvement project is implemented, a measurement system tracks expenses and revenues. Typically, quality costs should be integrated into the existing management cost-accounting system. Such a system is already in place and well understood by everyone. Managers have been trained and reinforced to use the cost-accounting system, which is often computerized. Therefore data can be assembled and distributed easily.

In the improvement project the operational department, rather than accounting, is responsible for assembling and distributing quality cost data. This offers the advantage that if the responsibility for assembling the data lies with the operational department, it will be more willing to implement any changes.

Interpret Results. The quality cost measurement system should measure hard and soft dollar benefits. Hard dollar benefits are those that can be controlled, measured, reduced, and managed. They result directly from lowering costs, improving operational effectiveness, or increasing market share. The quality cost measurement system cannot be sold to management in terms of vague and unsubstantiated numbers. Soft dollar benefits, not directly attributed to any quality improvement, usually will not justify a future quality project or the purchase of a piece of capital equipment.

Soft long-term costs, including lost sales due to customer dissatisfaction or liability exposure, are also very difficult to determine. Many of these numbers are not available in traditional cost-accounting systems, which capture only costs such as warranty work, scrap, rework, and recall.

Hard dollar benefits are well understood and manageable. For example, every operator and supervisor understands the significance of scrap, rework, and inspection costs. These are easy to measure. However, underutilized equipment and customer dissatisfaction are easy to understand but difficult to measure.

Once an improvement project is run, management and workers should be able to recognize the benefits of measuring quality costs. The benefits of the improvement project may however not be sufficiently obvious to justify a company-wide measurement effort, so another improvement project may be initiated.

TIME VALUE OF MONEY

Many top quality consultants, such as Joseph Juran and Philip Crosby, maintain that the cost of quality can be as high as 25% of gross sales. The cost of quality is the cost resulting from defective products and poor service, combined with the costs to prevent the defects from occurring.

Since the cost of quality is so high, Philip Crosby said, "quality is free," which is the title of a best-seller he wrote several years ago.* He meant that the returns of an effective quality program are higher than what it costs to implement the program.

For an overall quality program this is generally true. However, a quality improvement project must be based on its own merits. Each quality improvement proposal is evaluated in terms of its overall return and its return against other competing projects. Any quality decision requiring funds becomes an economic decision of internal investment for a firm, and it must be viewed as adding value or satisfying the final customer.

Some quality projects improve machines, processes, people, or systems, but have no economic merit. An example may illustrate this. A company recently expended a great deal of resources in a total quality management effort. People were trained, machines were purchased, substantial improvements were made. One year later the company went bankrupt. Popular wisdom dictated that quality improvements would make the company more competitive. Unfortunately the customer was not impressed by the internal improvements, which did not translate into a marketable product.

---

* P.B. Crosby, *Quality is Free* (McGraw-Hill, New York, 1979).

Furthermore, any quality improvement project must pay for itself. This type of economic analysis can be used in evaluating costs associated with overdesign, designing for process control, preventive maintenance, interchangeable design, and standardization.

The relationship between factors such as investment, interest, and time describes an important concept, called the **time value of money.** This concept can be explained easily by saying that a dollar received in the future is worth less than one available now.

Quality improvement requires the investment of money for property, plant, equipment, and training over a period of time. The time value of such investments and their returns are reflected in the evaluation of quality projects.

Money has earning power. An interest rate is usually stated on a per-year basis. Thus a 10% interest on $100 means a year from now the rental payment to use the $100 is $10.

Economic quality analysis is concerned with the evaluation of economic alternatives. These alternatives are described in terms of money going out (disbursements or investments) and money coming in (receipts or income). The amount and timing of money flows is important when economic alternatives are evaluated.

Interest

The time value of money depends on the magnitude of the interest. The term **interest** has several economic meanings, depending on the perspectives of the borrower, the lender, and the organization. From the borrower's perspective, interest is the rental amount charged by banks to borrow money. From a lender's perspective, interest is the gain obtained from an investment. From an organization's point of view, interest represents a firm's cost of doing business.

There is no standard, or "normal," interest rate. **Prime interest,** the rate at which the government loans money to prime lenders, is sometimes the benchmark on which many interest rates are based. Home mortgage, automobile, and credit-card interest is usually pegged at some higher rate than prime.

Interest rates fluctuate based on the supply and demand of money. If the demand for money exceeds the supply, the interest charged to obtain the money is higher than if the demand equals supply. If the supply of money exceeds the demand, the interest charged would be lower.

From an internal manufacturing or engineering manager's point of view, money will only be invested in a quality improvement project if it yields more than an investment of comparable risk or if its return is higher than a firm's cost of money. In addition, a risk factor has to be added to the cost of money. The risk is the probability that the borrower can repay the loan or, internally, that the projected inflows of cash will occur.

If a project is perceived as risky, interest rates will be higher than for a project with a guaranteed rate of return. Several examples may help to illustrate this from the lender's and borrower's perspectives. A lender can be thought of as an investor, a person who seeks to invest money in a company or project that will return more money at similar risks that competing investments. The borrower wants to select the lender who charges the least interest.

Types of Interest

The interest, or rental, rate of money is expressed as a percentage of the total amount of money that is paid for its use for a period of time. These time periods are called interest periods. For simplicity they are considered in this chapter to be 1 year. We discuss two types of interest, simple interest and compound interest.

Simple Interest. In **simple interest** the interest paid on a loan is related to the length of time the money was borrowed. In the language of economics, the interest paid for the use of the principal is directly related to the length of time the principal was borrowed. This statement, expressed by an equation, is

$$I = Pni$$

where I = interest earned
P = principal
n = period of time
i = interest rate, %

EXAMPLE $100 was borrowed at a simple interest rate of 10% per year. Calculate the interest obtained at the end of 1 year.

Solution
Given P = $100, n = 1 year, and i = 10% = 0.10,

$$I = Pni$$
$$= \$100 \times 1 \text{ year} \times 0.10 = \$10.00$$

The principal plus interest is $110, and is due in 1 year. If interest has to be calculated for a fraction of a year, the period of time n is divided by 360 days, which corresponds to 12 months at 30 days per month. For example, the above loan calculated for 2 months, or 60 days, for a principal of $100 is

$$I = Pni$$
$$= \$100 \times 60/360 \times 0.10 = \$1.66$$

Compound Interest. In the preceding example we showed how money has a time value. When at the end of each year, the total value of an account, that is, principal plus interest, is multiplied by the interest rate, this is called **compound interest.** Thus if the principal and interest are kept in an account for 2 years, at the end of this time, the principal plus interest of the first year are multiplied by the interest rate, and so on each following year. The interest is thus compounded annually.

To use our earlier example, if the $100 is kept in an account and interest is compounded annually at 10%, the $100 grows to $110 in one year. In the second year, at the same rate of 10%, the $110 becomes $121.

INTEREST FORMULAS

Cash-Flow Diagram

The flow of capital can be illustrated through a diagram called a **cash-flow diagram.** Cash flows are variously described as inflows/outflows, revenues/costs, receipts/ disbursements, or income/expenses. This diagram is especially useful for explaining complex positive and negative money flows over time. From the borrower's perspective, cash outflow is called negative cash flow and cash inflow is called positive cash flow. The cash-flow diagram shows inflow as a positive, upward arrow and outflow as a negative, downward arrow. The arrow's height may be proportional to the magnitude of the inflow or outflow. These arrows are placed on a horizontal line, which is a time line for the period of investment.

EXAMPLE   Over 4 years, a messenger service tracked the net payments of owning a truck. The truck's initial cost was $12,500, including title and 1 year's insurance. Each year the costs were tracked for insurance, gas, and incidentals. At the end of the first year, the cost of using the truck was $1500. At the end of the second year, costs increased to $2500 because of an additional cost of gas. In years 3 and 4, costs increased because of major repairs, including replacing an oil seal and repairing the transmission.

Solution

At the end of year 0	$12,500
At the end of year 1	1,500
At the end of year 2	2,500
At the end of year 3	4,000
At the end of year 4	4,000

The cash flows from the messenger's perspective are shown in Figure 10.3.

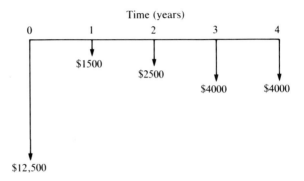

FIGURE 10.3
Cash-flow diagram.

EXAMPLE A cash-flow diagram is drawn differently depending on whether it is from the lender's or from the borrower's point of view. It is therefore essential to identify from whose perspective the diagram is being considered.

A bank lends a custom programming company $10,000 for a computer system. The company pays the bank $3000 each year. The cash flows for the lender and the borrower are shown in Figure 10.4.

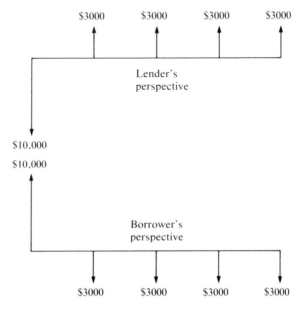

FIGURE 10.4
Cash-flow diagrams showing lender's and borrower's perspectives.

Formulas

In the earlier example the messenger service analyzed truck costs without paying attention to the cost of money. It would be very useful to be able to bring each future amount of money to the present, or vice versa. In the following sections we describe methods for calculating various types of cash flows using the time value of money, specifically the future value, the present value, and equal-payment series for future and present values as well as sinking funds.

The formulas use similar notation:

r = nominal interest rate per period

i = annual interest rate

n = number of annual interest periods

P = present value of money, principal

F = future value of money

A = uniform sum of money in each time period

Future Value. Suppose a quality analyst wants to know the **future value** of a single amount, $100, at 10% interest, compounded annually, in 5 years. Using the above notation, this is expressed as follows: $100 principal ($P$) is invested in the present with the earning power of 10% (i) per year. Principal and interest are compounded yearly. At the end of 5 (n) years, the amount accumulated (F) is calculated by the future value formula:

$$F = P(1 + i)^n$$

The factor in parentheses, to the right of the equal sign, is the single-payment compound factor, more commonly called the **future value factor.** It is abbreviated $\{F/P\ i,n\}$. This factor is composed of the interest rate i per period and the number of periods n between the present and the future. The abbreviation reads: the future value F is derived using the present value P, which is multiplied by the single-payment future value factor at an interest rate of i. To simplify these calculations, n is a whole number representing years and i is the interest rate per year.

The cash-flow diagram in Figure 10.5(a) illustrates this investment from the lender's perspective. It shows that payment is not made until an investment is terminated and interest is compounded. Earned interest is added to the principal at the end of each interest period.

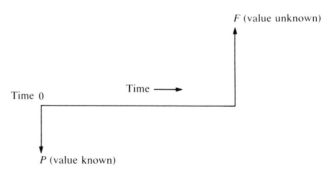

(a) Future value

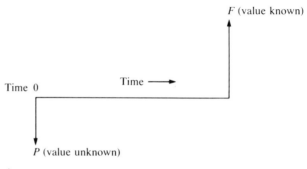

(b) Present value

FIGURE 10.5
Cash-flow diagrams of future and present values.

EXAMPLE   Referring to a previous example, $100 is invested for 5 years at 10% interest, compounded annually.

Solution
The compound amount at the end of the fifth year is

$$F = \$100(1 + 0.10)^5$$
$$= \$100(1.6105) = \$161.05$$

Using Appendix C, the calculation reduces to finding the appropriate factor from the table:

$$F = \$100 \; \{F/P \; 10, \; 5\}$$
$$F = \$100(1.6105) = \$161.05$$

Present Value. The quality analyst now wants to know the **present value** of a single future amount. Using the same notation and the preceding example, the future value of $161.05 (F), invested at 10% (i) per year for 5 years (n), is worth $100.00 in the present.

The present value equation is obtained by rearranging the future value formula and isolating P on one side as follows:

$$P = F \frac{1}{(1 + i)^n}$$

The fraction to the right is the single-payment present worth factor, also called the **present value factor.** This is expressed as $\{P/F \; i, \; n\}$. The cash-flow diagram for present value is shown in Figure 10.5(b). As you can see, the present value and future value factors are reciprocals of each other.

EXAMPLE   Find the present value P of a future amount F of the previous example.

Solution
Appropriate values can be substituted into the preceding equation:

$$P = \$161.05 \frac{1}{(1 + 0.10)^5}$$
$$= \$161.05(0.6209) = \$99.99$$
$$\approx \$100.00$$

This can also be calculated using Appendix C:

$$P = \$161.05\{P/F \; 10, \; 5\}$$
$$= \$161.05(0.6209)$$
$$\approx \$100.00$$

Future Value—Equal-Payment Series. Sometimes the quality analyst wants to know the future value that would accumulate from a series of equal payments occurring at the end of succeeding yearly periods. Expressed mathematically, to find the future value F of an equal-payment series A invested at interest rate i for n equal periods, the following formula applies:

$$F = A\frac{(1 + i)^n - 1}{i}$$

This calculation becomes complex quickly. It is easier to consult Appendix C, which lists the expression on the right, called the future value—equal-payment series factor and abbreviated $\{F/A\ i, n\}$. Figure 10.6(a) shows a cash-flow diagram for the future value of an equal-payment series.

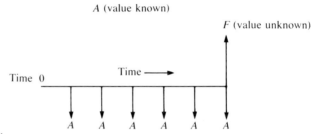

(a) Future value

(b) Present value

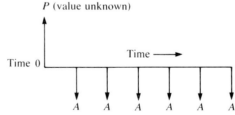

(c) Sinking fund

FIGURE 10.6
Cash-flow diagrams of equal-payment series.

EXAMPLE   Determine the future value of $500 deposited in $100 payments at the end of each of the next 5 years and earning 10% per year.

Solution

Time	Year-End Payment	Future Value
End of year 1	$100(1.10)^4	$146.41
End of year 2	$100(1.10)^3	$133.10
End of year 3	$100(1.10)^2	$121.00
End of year 4	$100(1.10)^1	$110.00
End of year 5	$100(1.10)^0	$100.00
	Total	$610.51

Using the preceding formula, the calculation becomes

$$F = \$100 \, \frac{(1 + 0.10)^5 - 1}{0.10}$$

$$= \$100(6.1051) = \$610.51$$

Using Appendix C and finding the appropriate factor, the calculation simplifies to

$$F = \$100 \, \{F/A \ 10, \ 5\}$$

$$= \$610.51$$

Present Value—Equal-Payment Series. A quality analyst wants to know the present value of a series of equal yearly payments. For example, companies invest into pension funds for their employees. When an employee wants to know whether she has sufficient money to retire, the company will calculate the present value of the retirement pension contributions.

To calculate the present value P of a series of n equal payments A at an interest rate i, the following formula applies:

$$P = A \, \frac{(1 + i)^n - 1}{i(1 + i)^n}$$

This formula looks complex, but it can be pictured easily in a cash-flow diagram, as illustrated in Figure 10.6(b).

EXAMPLE   A company knows that it will have profits of $100,000 at the end of each of 5 years and wants to invest the funds at 10% interest, compounded annually. This situation is called an **annuity,** and the company wants to know the present value of the annuity.

Solution
The present value of an annuity can be found by using the preceding equation:

$$P = A \, \frac{(1 + i)^n - 1}{i(1 + i)^n}$$

Given $A = \$100,000$, $n = 5$, and $i = 10\%$,

$$P = \$100,000 \,\frac{(1 + 0.10)^5 - 1}{0.10(1 + 0.10)^5}$$

$$= \$100,000(3.7908) = \$379,080$$

Using Appendix C and finding the appropriate factor, the calculation simplifies to

$$P = \$100,000 \,\{P/A \; 10, 5\}$$

$$= \$100,000(3.7908) = \$379,080$$

Sinking Fund—Equal-Payment Series. A company wants to purchase a new computer-controlled screw machine in 5 years because the existing machine will then have worn out. To purchase the machine, the company wants to set aside sufficient capital each year so that after 5 years, at a 10% compounding rate, it can buy the machine.

This purchasing method is very common in business and is called a **sinking fund.** It is used when a number n of equal periodic payments of value A are invested at an interest rate i to generate a known future amount of money F. The payment amount A is unknown.

The required amount of each payment is expressed by the factor $\{A/F, i, n\}$. This factor is used to find the required year-end payments necessary to accumulate a future amount F. The general formula for calculating a sinking fund is

$$A = F \,\frac{i}{(1 + i)^n - 1}$$

This process is called the sinking fund—equal-payment series and is illustrated in Figure 10.6(c).

EXAMPLE   The future value of a 5-year annuity earning interest at 10% compounded annually is $100,000. To have $100,000 at the end of 5 years, we want to know how much to invest in a sinking fund at the end of each year for the next 5 years at 10% interest compounded annually.

Solution
Using the preceding formula,

$$A = F \,\frac{i}{(1 + i)^n - 1}$$

We find

$$A = \$100,000 \,\frac{0.10}{(1 + 0.10)^5 - 1}$$

$$= \$100,000(0.16379) = \$16,379$$

Consulting Appendix C under $\{A/F, 10, 5\}$,

$$= \$100,000\{A/F\ 10,\ 5\}$$
$$= \$100,000(0.1638) = \$16,380$$

MEASURES OF EQUIVALENCE

Comparison of Alternatives

Quality management decisions often involve investment in property, buildings, equipment, materials, or training. These are important corporate assets, which a company commits with the anticipation of receiving some future gain. Traditional economic analysis treated quality costs as yearly expenditures, without much involvement with time, the cost of money, or returns. It was assumed that quality was a worthwhile endeavor, which should not be analyzed like other investment decisions. This is incorrect. Quality decisions must be evaluated like any other management decision, using the same type of analyses.

For example, a common quality management question is whether to make a product in house or buy it from a supplier. This decision becomes a question of weighing the merits of alternative investments and choosing the option offering the most advantages.

Using the earlier example of a screw machine, there are several alternatives. The old machine could be left alone, maintained, and repaired as necessary until it can no longer be used. Another decision would be to stop making the part altogether and buy the product from a supplier. Or if the machine is to be replaced by another one, should it be a fully automated, larger machine capable of performing multiple operations? Then there may be alternatives for using the funds allocated to the machine, which may generate a greater profit. All these options should be considered using the time value of money.

Analytical Tools

Any quality decision, whether a quality improvement project or a purchase of a piece of capital equipment, should minimize costs or generate more inflow than outflow. The interest formulas allow the quality analyst to compare and summarize differences between various types of projects. However, each project may have its own cash flows at different interest rates, time periods, and intervals. In order to perform this analysis, there must be a basis of equivalent comparison among investment alternatives. One method is to bring cash inflows and outflows into the present. Equivalence formulas bring the different cash flows to a common point or baseline so that all economic projects can be compared on an equivalent basis. For example, different cash flows for a machine tool may be brought to the present and compared. The investment with the highest return, all things being equal, is chosen.

Common methods for comparing economic projects are introduced in this section. If cash flows are being projected, it should be mentioned that they are based on projections, or forecasts, that may not come to pass since the past is only a rough indicator of future events. The following analytical tools are relatively easy to use: payback, simple return on investment, and net present value.

Payback. Funds are invested to obtain sufficient economic returns over a future period to justify the original investment. **Payback** is a simple measure, usually expressed in years, that relates the initial investment in a machine or improvement project to the average annual cash flow derived from the investment. Mathematically this is expressed as

$$\text{Payback} = \frac{\text{investment}}{\text{average annual cash flow}}$$

For example, a machine costs $100,000. The machine will generate $20,000 in yearly operating cash flow. The payback for this machine is a simple division of the investment divided by the average annual cash flow, which results in 5 years. This payback is the number of years required to pay off the initial investment. If a quality engineer knows the machine has a 5-year economic life, any use after the 5-year payback period does not entail further cost except for periodic maintenance.

The payback concept is easy to understand and to calculate. The simplicity of the payback method is balanced against its major drawback. Payback analysis does not consider the time value of money.

EXAMPLE The payback method can also be used to evaluate uneven cash flows. Given a series of uneven cash flows over a period of time, each flow is registered as positive inflow or negative outflow. When the accumulated cash flow becomes positive, this is the payback in years.

Solution

End of Year	Cash flow	Accumulated Sum of Cash Flows
1	−$10,000	−$10,000
2	6,000	−4,000
3	2,000	−2,000
4	2,000	0

In this example the payback is 4 years. In the first year $10,000 goes out, and in the succeeding years $6000, $2000, and $2000 come in. The accumulated sum of cash flows is zero at the end of the fourth year.

Simple Return on Investment. The **simple return on investment** (ROI) is another method for evaluating quality improvement projects. Return on investment expresses the economic worth of a project in terms of a percentage return on the initial investment. For example, the return on investment on the payback example in the preceding section is

$$\text{Return on investment} = \frac{\text{annual operating cash flow}}{\text{investment}}$$

$$= \frac{\$20,000}{\$100,000} = 20\%$$

The simple return on investment has the same advantages and drawbacks as the payback method. This method does not consider the time value of money.

Net Present Value. **Net present value** (NPV) is a method for evaluating economic alternatives that eliminates the problems of the two previous methods. This method considers the time value of money as well as differing cash flows.

In most economic quality analyses, the time value of money becomes a major consideration, especially in times when inflation decreases the value of a dollar each year. Also, it is important when a corporation has opportunities to invest funds in competing areas, for example, risk-free nominal-interest investments, such as government certificates of deposit, or high-return high-risk investments, such as precious metals or junk bonds.

The basic objective of net present value is to adjust outflows and inflows in terms of present value or time-adjusted dollars. The time value of money depends on the interest rate which, in turn, is based on the supply and demand of capital. This rate is the interest rate i in our calculations. Using this information, the quality analyst determines the present value of all outflows and inflows of the project. The result is a positive or negative number. A positive number indicates the project returns more than the initial investment compounded annually over its economic life. A negative number means the project is not achieving predetermined goals. A negative number usually kills the project unless it is required for safety, health, or some other important reasons. An example calculation of net present value is presented at the end of this chapter.

TAGUCHI QUALITY ECONOMICS

Dr. Genichi Taguchi, winner of four Deming prizes, developed a different approach to improve quality and lower costs by optimizing product design and manufacturing processes. His methods to reduce costs and improve quality, using advanced statistics called design of experiments, are beyond the scope of this text. In this section we introduce his important ideas only briefly.

Loss to Society

The essence of Taguchi's philosophy is that any loss of quality is a loss to society. This is a much broader definition of quality than we have discussed throughout this text. To Taguchi, loss may mean customer dissatisfaction, warranty costs, loss of market share, deterioration of safety and health, or failure to meet a customer's expectations. Taguchi believes if a product is produced that no one wants, scarce resources have gone into designing and manufacturing a product, which could have been better invested elsewhere. On the other hand, if a product fails in the field and results in a loss of life, this is another tragic loss to society.

Quality control as a science should strive to reduce the cost to society of poorly designed products and develop products, services, and projects that satisfy the customer. Taguchi believes customer satisfaction leads to increases in sales and subsequent increases in market share and income.

However, product quality must also be balanced against cost improvement. If a customer is given a choice between two equally priced products but of differing quality, the customer will always choose the higher perceived quality product. Likewise, if a customer is given a choice between two equal-quality products but of differing price, the customer will always choose the lower priced product. So in a global market where products can travel at will, people will purchase competitively priced quality products.

Constant Reduction of Variation

In the view of Taguchi, continuous quality improvement requires that there be a constant reduction in the variations, approaching the target value. First, quality characteristics, whether appearance, performance, maintenance, or service, are specified in measurable and realistic terms. Quality is specified in terms of a target and a specification spread.

Many believe quality is acceptable as long as a dimension is anywhere inside a specification spread. If a dimension is outside the specification spread, the product is clearly unacceptable. Taguchi believes the target dimension is the ideal, the optimum, and the ideal diminishes as characteristics move away from the target dimension.

Figure 10.7 illustrates this point. The upper specification limit (USL) is 2.005 in and the lower specification limit (LSL) is 1.995 in. The loss curve is a parabola whose bottom hits the target dimension of the specification, 2.000 in.

To Taguchi, this target is the ideal point for a process to attain consistently. Again referring to our discussion in Chapter 4, the secret to quality control is the ability to hit the target consistently and to limit any variation about this target. Taguchi reinforces this by saying that any movement from the target results in deteriorating quality until a process outside the specification limits produces products that are a loss to society.

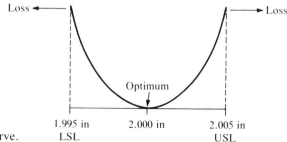

FIGURE 10.7
Taguchi loss curve.

SPOTLIGHT

A machine tool has been producing parts to a specified dimension for several years. The machine is getting sloppy, machine repeatability is low, repair costs are increasing, and downtime is slowing the production line. The machine is retained because it is too expensive to replace and the machine process is still capable of meeting specifications.

Engineering redesigns the existing product and specifies a tighter dimension of the quality characteristic being machined. The machine is no longer capable of producing products that conform to specifications. Specifications have been tightened to the point where inherent variation in the machine forces it to go outside the specification limits.

Manufacturing and quality management must decide whether to buy or make the product. If manufacturing room is limited or if people need additional training, then a buy decision is made and the part is purchased from a supplier. If the product is a key part and management wants to keep its production in house, the decision must be justified in terms of purchasing a new machine. To justify the expenditure of funds for a new machine, manufacturing considers all cash flows over the time the machine will be used.

The total cost of the machine over its useful life is considered, including initial cost, the net present values of yearly operating and maintenance costs, and scrap value. This type of analysis, also called life-cycle analysis, can be used for evaluating products, revenue-producing projects, and non-revenue-producing projects.

The buyer of the machine sees the machine tool as an investment from which the company wants a return greater than the combined initial cost, yearly operating and maintenance costs, installation costs, downtime, preventive maintenance, programming costs, spare parts, power costs, operator pay, supervision salary, and overhead costs. Yearly outflows are subtracted from inflows and if inflows are sufficiently greater than outflows, the machine is purchased. These are traditional cost-accounting data, which are available in the plant chart of accounts.

Initial Costs

	$200,000	Machine
	10,000	Installation
	10,000	Programming
	5,000	Operator training
Total	$225,000	

First-Year costs

	$30,000	Operator wages
	10,000	Raw material
	7,000	Total overhead
	5,000	Preventive maintenance
	3,000	Spare parts
Total	$55,000	

Second-Year costs

$30,000	Operator wages
15,000	Raw material
7,000	Total overhead
5,000	Preventive maintenance
3,000	Spare parts
Total $60,000	

If the company bought parts from a supplier, the parts would cost $100,000 a year. In the first year the net cash benefit of purchasing the machine is $45,000 ($100,000 − $55,000). From experience with similar types of machines, manufacturing knows that when the machine is fitted into the just-in-time production line in the second year, production will increase about 50%. All costs will remain the same except the cost of raw material. If production increases 50%, the cost of raw material increases to $15,000. Thus in the second year the net cash flow is $90,000 ($150,000 − $60,000). In the third and fourth years, net cash flows remain the same.

In years 5 through 7, the machine starts wearing out. Production decreases, downtime increases, and repairs increase. By this time the net cash flow decreases to $25,000 ($120,000 − $95,000).

Years 5–7

$30,000	Operator wages
15,000	Raw material
15,000	Downtime
13,000	Repair
10,000	Spare parts
7,000	Total overhead
5,000	Preventive maintenance
Total $95,000	

The company decides to invest in a newer, more advanced machine. The old machine is sold and $20,000 is obtained. This amount is reflected in the final (year 7) net cash flow, which becomes $45,000 ($25,000 + $20,000).

Suppose three machines are evaluated similarly and each machine has a different initial cost, net yearly cash flow, and ending scrap value. Looking at the cash flows, there is no easy way to determine which is the best investment. We must make the investment options equivalent so there can be a basis for comparison. Here the best method is the net present value method because it incorporates the time value of money. Each yearly cash flow is different, so each year must be brought to the present. Let us assume the interest rate is 10% for each of the machines. Then the cash flows for years 0–7, machines 1–3, are as follows:

Year	Machine 1	Machine 2	Machine 3
0	−$225,000	−$200,000	−$280,000
1	45,000	35,000	55,000
2	90,000	65,000	95,000
3	90,000	65,000	95,000
4	90,000	65,000	95,000
5	25,000	20,000	30,000
6	25,000	20,000	30,000
7	45,000	10,000	55,000

1. Machine 1:

Present Value = $225,000 \{P/F\ 10,\ 0\} + \$45,000\ \{P/F\ 10,\ 1\} +$
$\$90,000\ \{P/F\ 10,\ 2\} + \$90,000\ \{P/F\ 10,\ 3\} +$
$\$90,000\ \{P/F\ 10,\ 4\} + \$25,000\ \{P/F\ 10,\ 5\} +$
$\$25,000\ \{P/F\ 10,\ 6\} + \$45,000\ \{P/F\ 10,\ 7\}$

$PV_1 = -225,000 + 40,910 + 74,385 + 67,617 + 61,470$
$+ 15,523 + 14,113 + 23,094$

$= \$72,112$

2. Machine 2:

$$PV_2 = \$7610$$

3. Machine 3:

$$PV_3 = \$48,564$$

Machine 1 has the highest positive net present value, so it is the best investment. We could calculate the future value of each machine. However, the present value offers a major advantage for establishing a suitable basis for comparison. The present value calculates the equivalent value of the cash flows into a single index at the time of purchase.

SUMMARY

☐ Investment decision making is not covered in quality assurance and control books. Quality projects must be evaluated in a similar way as other investment alternatives. Quality management decisions involve the deployment of scarce corporate resources. These compete against other investment alternatives. If a quality project is to be implemented, it must offer a greater reward at a similar risk than competing projects.

☐ A quality cost system is an important element of a CWQM program. A reliable measurement of costs is essential to setting quality objectives, prioritizing objectives, establishing accountabilities, and funding improvement projects.

☐ A quality cost measurement program follows these sequential steps: obtain management support, identify improvement project, determine baseline costs, implement quality measurement systems, and interpret results.

☐ Traditional quality costs are broken down into

two components: conformance and nonconformance costs. Conformance costs are incurred as a result of making products conform to specifications or services conform to procedures. Nonconformance costs are incurred as a result of not complying with specifications or procedures.

☐ Conformance costs are broken down into prevention and appraisal. Nonconformance costs are broken down into internal and external failure costs. These costs can be displayed on cost curves. The total cost curve is the sum of nonconformance and conformance curves. The minimum point on the total cost curve is the optimum balance between nonconformance and conformance costs.

☐ Reliability and maintainability are graphed similarly as the conforming and nonconforming curves. If both curves are summed, the result is the total reliability cost curve. The optimum cost on this curve is somewhere to the right of the lowest point on the curve. Usually, increased reliability is worth the additional investment in costs.

☐ Any quality investment over time should be evaluated in terms of the cost of money. The time value of money is represented by the interest rate. Interest can be defined in terms of borrower, lender, or organization. A useful definition is to consider interest as the rental rate of money. Interest should be incorporated into any cost/benefit analysis.

☐ Investment analysis is concerned with the analysis of economic alternatives. This is sometimes called capital budgeting.

☐ Economic analysis is concerned with periods of time, usually in increments of 1 year or some part of a year. These are called interest periods. Two types of interest are used, simple and compound interest. Simple interest is the amount of interest paid for a specific length of time. Simple interest does not consider the time value of money. Compound interest explains how money compounds itself over time. Almost all investment decisions dealing with quality use compound interest.

☐ Interest formulas use compound interest. The formulas consist of factors that relate the time value of money to different investment alternatives. In this chapter we discussed the following factors: future value, present value, and equal-payment series of future and present values as well as sinking funds.

☐ Using these factors, investment decisions may be reached by comparing the cash flows of various alternative investments. In order to make investment alternatives comparable, three common methods of equivalence are introduced: payback, simple return on investment, and net present value. The net present value is preferred because it incorporates the time value of money.

☐ Taguchi has developed a novel approach to quality costs. He defines the loss of quality as the loss to society of not attaining quality. Taguchi believes customer dissatisfaction or product failure leads to a loss of sales and a subsequent loss of market share.

KEY TERMS

annuity   Payment of fixed sum of money at regular periods of time.

appraisal costs   Costs incurred to assess and measure the existing state of quality.

cash-flow diagram   Diagram illustrating the inflows and outflows of cash.

compound interest   Interest relating time, interest, and principal amount over successive periods of time.

conformance costs   Costs incurred in attempting to conform to product specifications or service requirements.

external failure costs   Costs incurred as a result of poor supplied material, warranty service, or product liability.

future value   Future worth of money.

future value factor   Factor to calculate the future amount of principal and interest compounded annually.

interest   Rental rate for money.

internal failure costs   Costs incurred as a result of rework, scrap, or product testing.

net present value (NPV)   Measure of equivalence; net value of cash flows brought to the present.

nonconformance costs   Costs incurred in not being able to conform to product specifications or service requirements.

payback   Measure of equivalence; sum invested divided by average annual cash flow.

present value   Value of money brought to the present.

present value factor   Factor to bring future value

to the present if principal and interest are compounded.

prevention costs   Costs incurred to prevent defects, flaws, or nonconformances from occurring.

prime interest   Interest rate at which government loans money to banks.

quality costs   Cost system consisting of conformance and nonconformance costs.

simple interest   Interest related to the time the principal was borrowed.

simple return on investment (ROI)   Measure of equivalence; annual cash flow divided by sum invested.

sinking fund   Year-end payments necessary to accumulate a specified future amount.

Taguchi, Dr. Genichi   Japanese authority on quality management; defines quality deterioration as loss to society.

time value of money   Earning power of money; in other words, a dollar today is worth more than a dollar tomorrow.

QUESTIONS

1. Why is it important that investment decision making be taught in a textbook on quality?

2. What are the purposes of a quality cost measurement system?

3. Explain the steps for implementing a quality cost measurement system.

4. Should accounting, quality, or some other part of the organization be responsible for capturing quality costs?

5. Define conformance and nonconformance costs.

6. What elements comprise internal and external failure costs?

7. Draw a set of typical cost curves and label each appropriately.

8. Where is the theoretical minimum cost on the total cost curve?

9. Draw a set of typical reliability and maintainability cost curves.

10. Is quality free? Discuss.

11. How is the definition of interest different from the perspectives of the lender and the borrower?

12. The time value of money should be considered when evaluating different economic alternatives. Discuss.

13. What is the difference between simple and compound interest?

14. When would a sinking fund—equal-payment series be used?

15. Why is it necessary to find an equivalent method to compare alternative investments?

16. When should payback and simple return on investment be used?

17. What is the main advantage of the net present value technique of analysis?

18. What does Taguchi mean when he defines quality as a "loss to society"?

PROBLEMS

1. What future values will be accumulated in the following investments? Draw a cash-flow diagram for each problem.
   a. $10,000 in 5 years at 10% compounded annually.
   b. $4500 in 8 years at 4% compounded annually.
   c. $3200 in 3 years at 6% compounded annually.

2. What future values will be accumulated in the following investments? Draw a cash-flow diagram for each problem.
   a. $5500 in 7 years at 5% compounded annually.
   b. $7500 in 4 years at 8% compounded annually.
   c. $12,000 in 5 years at 10% compounded annually.

3. What are the present values of the following amounts? Draw a cash-flow diagram for each problem.
   a. $10,000 in 4 years at 4% compounded annually.
   b. $2300 in 8 years at 8% compounded annually.
   c. $4000 in 5 years at 6% compounded annually.

4. What are the present values of the following amounts? Draw a cash-flow diagram for each problem.
   a. $15,000 in 6 years at 8% compounded annually.
   b. $2600 in 4 years at 7% compounded annually.
   c. $1800 in 8 years at 6% compounded annually.

5. What are the future values of the following series of payments? Draw a cash-flow diagram for each problem.
   a. $200 yearly for 4 years at 5% compounded annually.
   b. $2500 yearly for 8 years at 8% compounded annually.
   c. $7800 yearly for 13 years at 5% compounded annually.

6. What are the future values of the following series of payments? Draw a cash-flow diagram for each problem.
   a. $1000 yearly for 7 years at 9% compounded annually.
   b. $1600 yearly for 6 years at 3% compounded annually.
   c. $6800 yearly for 4 years at 7% compounded annually.

7. What are the present values of the series of payments in Problem 5?

8. What are the present values of the series of payments in Problem 6?

9. What series of equal payments would repay the following present values? Draw a cash-flow diagram for each problem.
   a. $4000 in 3 years at 8% compounded annually.
   b. $5600 in 5 years at 9% compounded annually.
   c. $7800 in 7 years at 3% compounded annually.

10. What series of equal payments would repay the following present values? Draw a cash-flow diagram for each problem.
    a. $3400 in 8 years at 8% compounded annually.
    b. $8900 in 5 years at 4% compounded annually.
    c. $5500 in 3 years at 7% compounded annually.

11. What series of equal payments into a sinking fund would result in the following amounts? Draw a cash-flow diagram for each problem.

    a. $10,000 in 4 years at 7% compounded annually.

    b. $4000 in 7 years at 8% compounded annually.

    c. $2500 in 5 years at 5% compounded annually.

12. What series of equal payments into a sinking fund would result in the following amounts? Draw a cash-flow diagram for each problem.

    a. $3200 in 3 years at 6% compounded annually.

    b. $8700 in 5 years at 8% compounded annually.

    c. $4000 in 4 years at 10% compounded annually.

13. How long will it take for an investment to double when it is invested at 8%, compounded annually?

Appendices

APPENDIX A Areas under the normal curve[a]

$\frac{X_i - \mu}{\sigma}$	0.09	0.08	0.07	0.06	0.05	0.04	0.03	0.02	0.01	0.00
−3.5	0.00017	0.00017	0.00018	0.00019	0.00019	0.00020	0.00021	0.00022	0.00022	0.00023
−3.4	0.00024	0.00025	0.00026	0.00027	0.00028	0.00029	0.00030	0.00031	0.00033	0.00034
−3.3	0.00035	0.00036	0.00038	0.00039	0.00040	0.00042	0.00043	0.00045	0.00047	0.00048
−3.2	0.00050	0.00052	0.00054	0.00056	0.00058	0.00060	0.00062	0.00064	0.00066	0.00069
−3.1	0.00071	0.00074	0.00076	0.00079	0.00082	0.00085	0.00087	0.00090	0 00094	0.00097
−3.0	0.00100	0.00104	0.00107	0.00111	0.00114	0.00118	0.00122	0.00126	0.00131	0.00135
−2.9	0.0014	0.0014	0.0015	0.0015	0.0016	0.0016	0.0017	0.0017	0.0018	0.0019
−2.8	0.0019	0.0020	0.0021	0.0021	0.0022	0.0023	0.0023	0.0024	0.0025	0.0026
−2.7	0.0026	0.0027	0.0028	0.0029	0.0030	0.0031	0.0032	0.0033	0.0034	0.0035
−2.6	0.0036	0.0037	0.0038	0.0039	0.0040	0.0041	0.0043	0.0044	0.0045	0.0047
−2.5	0.0048	0.0049	0.0051	0.0052	0.0054	0.0055	0.0057	0.0059	0.0060	0.0062
−2.4	0.0064	0.0066	0.0068	0.0069	0.0071	0.0073	0.0075	0.0078	0.0080	0.0082
−2.3	0.0084	0.0087	0.0089	0.0091	0.0094	0.0096	0.0099	0.0102	0.0104	0.0107
−2.2	0.0110	0.0113	0.0116	0.0119	0.0122	0.0125	0.0129	0.0132	0.0136	0.0139
−2.1	0.0143	0.0146	0.0150	0.0154	0.0158	0.0162	0.0166	0.0170	0.0174	0.0179
−2.0	0.0183	0.0188	0.0192	0.0197	0.0202	0.0207	0.0212	0.0217	0.0222	0.0228
−1.9	0.0233	0.0239	0.0244	0.0250	0.0256	0.0262	0.0268	0.0274	0.0281	0.0287
−1.8	0.0294	0.0301	0.0307	0.0314	0.0322	0.0329	0.0336	0.0344	0.0351	0.0359
−1.7	0.0367	0.0375	0.0384	0.0392	0.0401	0.0409	0.0418	0.0427	0.0436	0.0446
−1.6	0.0455	0.0465	0.0475	0.0485	0.0495	0.0505	0.0516	0.0526	0.0537	0.0548
−1.5	0.0559	0.0571	0.0582	0.0594	0.0606	0.0618	0.0630	0.0643	0.0655	0.0668
−1.4	0.0681	0.0694	0.0708	0.0721	0.0735	0.0749	0.0764	0.0778	0.0793	0.0808
−1.3	0.0823	0.0838	0.0853	0.0869	0.0885	0.0901	0.0918	0.0934	0.0951	0.0968
−1.2	0.0895	0.1003	0.1020	0.1038	0.1057	0.1075	0.1093	0.1112	0.1131	0.1151
−1.1	0.1170	0.1190	0.1210	0.1230	0.1251	0.1271	0.1292	0.1314	0.1335	0.1357
−1.0	0.1379	0.1401	0.1423	0.1446	0.1469	0.1492	0.1515	0.1539	0.1562	0.1587
−0.9	0.1611	0.1635	0.1660	0.1685	0.1711	0.1736	0.1762	0.1788	0.1814	0.1841
−0.8	0.1867	0.1894	0.1922	0.1949	0.1977	0.2005	0.2033	0.2061	0.2090	0.2119
−0.7	0.2148	0.2177	0.2207	0.2236	0.2266	0.2297	0.2327	0.2358	0.2389	0.2420
−0.6	0.2451	0.2483	0.2514	0.2546	0.2578	0.2611	0.2643	0.2676	0.2709	0.2743
−0.5	0.2776	0.2810	0.2843	0.2877	0.2912	0.2946	0.2981	0.3015	0.3050	0.3085
−0.4	0.3121	0.3156	0.3192	0.3228	0.3264	0.3300	0.3336	0.3372	0.3409	0.3446
−0.3	0.3483	0.3520	0.3557	0.3594	0.3632	0.3669	0.3707	0.3745	0.3783	0.3821
−0.2	0.3859	0.3897	0.3936	0.3974	0.4013	0.4052	0.4090	0.4129	0.4168	0.4207
−0.1	0.4247	0.4286	0.4325	0.4364	0.4404	0.4443	0.4483	0.4522	0.4562	0.4602
−0.0	0.4641	0.4681	0.4721	0.4761	0.4801	0.4840	0.4880	0.4920	0.4960	0.5000

[a] Proportion of total area under the curve that is under the portion of the curve from $-\infty$ to $(X_i - \mu)/\sigma$ (X_i represents any desired value of the variable X).

Source: Dale H. Besterfield, *Quality Control*, 3e, © 1990, pp. 420–421. Reprinted by permission of Prentice-Hall, Inc., Englewood Cliffs, New Jersey.

$\dfrac{X_i - \mu}{\sigma}$	0.00	0.01	0.02	0.03	0.04	0.05	0.06	0.07	0.08	0.09
+0.0	0.5000	0.5040	0.5080	0.5120	0.5160	0.5199	0.5239	0.5279	0.5319	0.5359
+0.1	0.5398	0.5438	0.5478	0.5517	0.5557	0.5596	0.5636	0.5675	0.5714	0.5753
+0.2	0.5793	0.5832	0.5871	0.5910	0.5948	0.5987	0.6026	0.6064	0.6103	0.6141
+0.3	0.6179	0.6217	0.6255	0.6293	0.6331	0.6368	0.6406	0.6443	0.6480	0.6517
+0.4	0.6554	0.6591	0.6628	0.6664	0.6700	0.6736	0.6772	0.6808	0.6844	0.6879
+0.5	0.6915	0.6950	0.6985	0.7019	0.7054	0.7088	0.7123	0.7157	0.7190	0.7224
+0.6	0.7257	0.7291	0.7324	0.7357	0.7389	0.7422	0.7454	0.7486	0.7517	0.7549
+0.7	0.7580	0.7611	0.7642	0.7673	0.7704	0.7734	0.7764	0.7794	0.7823	0.7852
+0.8	0.7881	0.7910	0.7939	0.7967	0.7995	0.8023	0.8051	0.8079	0.8106	0.8133
+0.9	0.8159	0.8186	0.8212	0.8238	0.8264	0.8289	0.8315	0.8340	0.8365	0.8389
+1.0	0.8413	0.8438	0.8461	0.8485	0.8508	0.8531	0.8554	0.8577	0.8599	0.8621
+1.1	0.8643	0.8665	0.8686	0.8708	0.8729	0.8749	0.8770	0.8790	0.8810	0.8830
+1.2	0.8849	0.8869	0.8888	0.8907	0.8925	0.8944	0.8962	0.8980	0.8997	0.9015
+1.3	0.9032	0.9049	0.9066	0.9082	0.9099	0.9115	0.9131	0.9147	0.9162	0.9177
+1.4	0.9192	0.9207	0.9222	0.9236	0.9251	0.9265	0.9279	0.9292	0.9306	0.9319
+1.5	0.9332	0.9345	0.9357	0.9370	0.9382	0.9394	0.9406	0.9418	0.9429	0.9441
+1.6	0.9452	0.9463	0.9474	0.9484	0.9495	0.9505	0.9515	0.9525	0.9535	0.9545
+1.7	0.9554	0.9564	0.9573	0.9582	0.9591	0.9599	0.9608	0.9616	0.9625	0.9633
+1.8	0.9641	0.9649	0.9656	0.9664	0.9671	0.9678	0.9686	0.9693	0.9699	0.9706
+1.9	0.9713	0.9719	0.9726	0.9732	0.9738	0.9744	0.9750	0.9756	0.9761	0.9767
+2.0	0.9773	0.9778	0.9783	0.9788	0.9793	0.9798	0.9803	0.9808	0.9812	0.9817
+2.1	0.9821	0.9826	0.9830	0.9834	0.9838	0.9842	0.9846	0.9850	0.9854	0.9857
+2.2	0.9861	0.9864	0.9868	0.9871	0.9875	0.9878	0.9881	0.9884	0.9887	0.9890
+2.3	0.9893	0.9896	0.9898	0.9901	0.9904	0.9906	0.9909	0.9911	0.9913	0.9916
+2.4	0.9918	0.9920	0.9922	0.9925	0.9927	0.9929	0.9931	0.9932	0.9934	0.9936
+2.5	0.9938	0.9940	0.9941	0.9943	0.9945	0.9946	0.9948	0.9949	0.9951	0.9952
+2.6	0.9953	0.9955	0.9956	0.9957	0.9959	0.9960	0.9961	0.9962	0.9963	0.9964
+2.7	0.9965	0.9966	0.9967	0.9968	0.9969	0.9970	0.9971	0.9972	0.9973	0.9974
+2.8	0.9974	0.9975	0.9976	0.9977	0.9977	0.9978	0.9979	0.9979	0.9980	0.9981
+2.9	0.9981	0.9982	0.9983	0.9983	0.9984	0.9984	0.9985	0.9985	0.9986	0.9986
+3.0	0.99865	0.99869	0.99874	0.99878	0.99882	0.99886	0.99889	0.99893	0.99896	0.99900
+3.1	0.99903	0.99906	0.99910	0.99913	0.99915	0.99918	0.99921	0.99924	0.99926	0.99929
+3.2	0.99931	0.99934	0.99936	0.99938	0.99940	0.99942	0.99944	0.99946	0.99948	0.99950
+3.3	0.99952	0.99953	0.99955	0.99957	0.99958	0.99960	0.99961	0.99962	0.99964	0.99965
+3.4	0.99966	0.99967	0.99969	0.99970	0.99971	0.99972	0.99973	0.99974	0.99975	0.99976
+3.5	0.99977	0.99978	0.99978	0.99979	0.99980	0.99981	0.99981	0.99982	0.99983	0.99983

APPENDIX B The Poisson Distribution $P(c) = (np_0/c!)e^{-np_0}$ (Cumulative values are in parentheses)

c	np_0 0.1		0.2		0.3		0.4		0.5	
0	0.905	(0.905)	0.819	(0.819)	0.741	(0.741)	0.670	(0.670)	0.607	(0.607)
1	0.091	(0.996)	0.164	(0.983)	0.222	(0.963)	0.268	(0.938)	0.303	(0.910)
2	0.004	(1.000)	0.016	(0.999)	0.033	(0.996)	0.054	(0.992)	0.076	(0.986)
3			0.010	(1.000)	0.004	(1.000)	0.007	(0.999)	0.013	(0.999)
4							0.001	(1.000)	0.001	(1.000)

c	np_0 0.6		0.7		0.8		0.9		1.0	
0	0.549	(0.549)	0.497	(0.497)	0.449	(0.449)	0.406	(0.406)	0.368	(0.368)
1	0.329	(0.878)	0.349	(0.845)	0.359	(0.808)	0.366	(0.772)	0.368	(0.736)
2	0.099	(0.977)	0.122	(0.967)	0.144	(0.952)	0.166	(0.938)	0.184	(0.920)
3	0.020	(0.997)	0.028	(0.995)	0.039	(0.991)	0.049	(0.987)	0.061	(0.981)
4	0.003	(1.000)	0.005	(1.000)	0.008	(0.999)	0.011	(0.998)	0.016	(0.997)
5					0.001	(1.000)	0.002	(1.000)	0.003	(1.000)

c	np_0 1.1		1.2		1.3		1.4		1.5	
0	0.333	(0.333)	0.301	(0.301)	0.273	(0.273)	0.247	(0.247)	0.223	(0.223)
1	0.366	(0.699)	0.361	(0.662)	0.354	(0.627)	0.345	(0.592)	0.335	(0.558)
2	0.201	(0.900)	0.217	(0.879)	0.230	(0.857)	0.242	(0.834)	0.251	(0.809)
3	0.074	(0.974)	0.087	(0.966)	0.100	(0.957)	0.113	(0.947)	0.126	(0.935)
4	0.021	(0.995)	0.026	(0.992)	0.032	(0.989)	0.039	(0.986)	0.047	(0.982)
5	0.004	(0.999)	0.007	(0.999)	0.009	(0.998)	0.011	(0.997)	0.014	(0.996)
6	0.001	(1.000)	0.001	(1.000)	0.002	(1.000)	0.003	(1.000)	0.004	(1.000)

c	np_0 1.6		1.7		1.8		1.9		2.0	
0	0.202	(0.202)	0.183	(0.183)	0.165	(0.165)	0.150	(0.150)	0.135	(0.135)
1	0.323	(0.525)	0.311	(0.494)	0.298	(0.463)	0.284	(0.434)	0.271	(0.406)
2	0.258	(0.783)	0.264	(0.758)	0.268	(0.731)	0.270	(0.704)	0.271	(0.677)
3	0.138	(0.921)	0.149	(0.907)	0.161	(0.892)	0.171	(0.875)	0.180	(0.857)
4	0 055	(0.976)	0.064	(0.971)	0.072	(0.964)	0.081	(0.956)	0.090	(0.947)
5	0.018	(0.994)	0.022	(0 993)	0.026	(0.990)	0.031	(0.987)	0.036	(0.983)
6	0.005	(0.999)	0.006	(0.999)	0.008	(0.998)	0.010	(0.997)	0.012	(0.995)
7	0.001	(1.000)	0.001	(1.000)	0.002	(1.000)	0.003	(1.000)	0.004	(0.999)
8									0.001	(1.000)

Source: Dale H. Besterfield, *Quality Control*, 3e, © 1990, pp. 423–427. Reprinted by permission of Prentice-Hall, Inc., Englewood Cliffs, New Jersey.

np_0 c	2.1		2.2		2.3		2.4		2.5	
0	0.123	(0.123)	0.111	(0.111)	0.100	(0.100)	0.091	(0.091)	0.082	(0.082)
1	0.257	(0.380)	0.244	(0.355)	0.231	(0.331)	0.218	(0.309)	0.205	(0.287)
2	0.270	(0.650)	0.268	(0.623)	0.265	(0.596)	0.261	(0.570)	0.256	(0.543)
3	0.189	(0.839)	0.197	(0.820)	0.203	(0.799)	0.209	(0.779)	0.214	(0.757)
4	0.099	(0.938)	0.108	(0.928)	0.117	(0.916)	0.125	(0.904)	0.134	(0.891)
5	0.042	(0.980)	0.048	(0.976)	0.054	(0.970)	0.060	(0.964)	0.067	(0.958)
6	0.015	(0.995)	0.017	(0.993)	0.021	(0.991)	0.024	(0.988)	0.028	(0.986)
7	0.004	(0.999)	0.005	(0.998)	0.007	(0.998)	0.008	(0.996)	0.010	(0.996)
8	0.001	(1.000)	0.002	(1.000)	0.002	(1.000)	0.003	(0.999)	0.003	(0.999)
9							0.001	(1.000)	0.001	(1.000)

np_0 c	2.6		2.7		2.8		2.9		3.0	
0	0.074	(0.074)	0.067	(0.067)	0.061	(0.061)	0.055	(0.055)	0.050	(0.050)
1	0.193	(0.267)	0.182	(0.249)	0.170	(0.231)	0.160	(0.215)	0.149	(0.199)
2	0.251	(0.518)	0.245	(0.494)	0.238	(0.469)	0.231	(0.446)	0.224	(0.423)
3	0.218	(0.736)	0.221	(0.715)	0.223	(0.692)	0.224	(0.670)	0.224	(0.647)
4	0.141	(0.877)	0.149	(0.864)	0.156	(0.848)	0.162	(0.832)	0.168	(0.815)
5	0.074	(0.951)	0.080	(0.944)	0.087	(0.935)	0.094	(0.926)	0.101	(0.916)
6	0.032	(0.983)	0.036	(0.980)	0.041	(0.976)	0.045	(0.971)	0.050	(0.966)
7	0.012	(0.995)	0.014	(0.994)	0.016	(0.992)	0.019	(0.990)	0.022	(0.988)
8	0.004	(0.999)	0.005	(0.999)	0.006	(0.998)	0.007	(0.997)	0.008	(0.996)
9	0.001	(1.000)	0.001	(1.000)	0.002	(1.000)	0.002	(0.999)	0.003	(0.999)
10							0.001	(1.000)	0.001	(1.000)

np_0 c	3.1		3.2		3.3		3.4		3.5	
0	0.045	(0.045)	0.041	(0.041)	0.037	(0.037)	0.033	(0.033)	0.030	(0.030)
1	0.140	(0.185)	0.130	(0.171)	0.122	(0.159)	0.113	(0.146)	0.106	(0.136)
2	0.216	(0.401)	0.209	(0.380)	0.201	(0.360)	0.193	(0.339)	0.185	(0.321)
3	0.224	(0.625)	0.223	(0.603)	0.222	(0.582)	0.219	(0.558)	0.216	(0.537)
4	0.173	(0.798)	0.178	(0.781)	0.182	(0.764)	0.186	(0.744)	0.189	(0.726)
5	0.107	(0.905)	0.114	(0.895)	0.120	(0.884)	0.126	(0.870)	0.132	(0.858)
6	0.056	(0.961)	0.061	(0.956)	0.066	(0.950)	0.071	(0.941)	0.077	(0.935)
7	0.025	(0.986)	0.028	(0.984)	0.031	(0.981)	0.035	(0.976)	0.038	(0.973)
8	0.010	(0.996)	0.011	(0.995)	0.012	(0.993)	0.015	(0.991)	0.017	(0.990)
9	0.003	(0.999)	0.004	(0.999)	0.005	(0.998)	0.006	(0.997)	0.007	(0.997)
10	0.001	(1.000)	0.001	(1.000)	0.002	(1.000)	0.002	(0.999)	0.002	(0.999)
11							0.001	(1.000)	0.001	(1.000)

c	np_0 3.6		3.7		3.8		3.9		4.0	
0	0.027	(0.027)	0.025	(0.025)	0.022	(0.022)	0.020	(0.020)	0.018	(0.018)
1	0.098	(0.125)	0.091	(0.116)	0.085	(0.107)	0.079	(0.099)	0.073	(0.091)
2	0.177	(0.302)	0.169	(0.285)	0.161	(0.268)	0.154	(0.253)	0.147	(0.238)
3	0.213	(0.515)	0.209	(0.494)	0.205	(0.473)	0.200	(0.453)	0.195	(0.433)
4	0.191	(0.706)	0.193	(0.687)	0.194	(0.667)	0.195	(0.648)	0.195	(0.628)
5	0.138	(0.844)	0.143	(0.830)	0.148	(0.815)	0.152	(0.800)	0.157	(0.785)
6	0.083	(0.927)	0.088	(0.918)	0.094	(0.909)	0.099	(0.899)	0.104	(0.889)
7	0.042	(0.969)	0.047	(0.965)	0.051	(0.960)	0.055	(0.954)	0.060	(0.949)
8	0.019	(0.988)	0.022	(0.987)	0.024	(0.984)	0.027	(0.981)	0.030	(0.979)
9	0.008	(0.996)	0.009	(0.996)	0.010	(0.994)	0.012	(0.993)	0.013	(0.992)
10	0.003	(0.999)	0.003	(0.999)	0.004	(0.998)	0.004	(0.997)	0.005	(0.997)
11	0.001	(1.000)	0.001	(1.000)	0.001	(0.999)	0.002	(0.999)	0.002	(0.999)
12					0.001	(1.000)	0.001	(1.000)	0.001	(1.000)

c	np_0 4.1		4.2		4.3		4.4		4.5	
0	0.017	(0.017)	0.015	(0.015)	0.014	(0.014)	0.012	(0.012)	0.011	(0.011)
1	0.068	(0.085)	0.063	(0.078)	0.058	(0.072)	0.054	(0.066)	0.050	(0.061)
2	0.139	(0.224)	0.132	(0.210)	0.126	(0.198)	0.119	(0.185)	0.113	(0.174)
3	0.190	(0.414)	0.185	(0.395)	0.180	(0.378)	0.174	(0.359)	0.169	(0.343)
4	0.195	(0.609)	0.195	(0.590)	0.193	(0.571)	0.192	(0.551)	0.190	(0.533)
5	0.160	(0.769)	0.163	(0.753)	0.166	(0.737)	0.169	(0.720)	0.171	(0.704)
6	0.110	(0.879)	0.114	(0.867)	0.119	(0.856)	0.124	(0.844)	0.128	(0.832)
7	0.064	(0.943)	0.069	(0.936)	0.073	(0.929)	0.078	(0.922)	0.082	(0.914)
8	0.033	(0.976)	0.036	(0.972)	0.040	(0.969)	0.043	(0.965)	0.046	(0.960)
9	0.015	(0.991)	0.017	(0.989)	0.019	(0.988)	0.021	(0.986)	0.023	(0.983)
10	0.006	(0.997)	0.007	(0.996)	0.008	(0.996)	0.009	(0.995)	0.011	(0.994)
11	0.002	(0.999)	0.003	(0.999)	0.003	(0.999)	0.004	(0.999)	0.004	(0.998)
12	0.001	(1.000)	0.001	(1.000)	0.001	(1.000)	0.001	(1.000)	0.001	(0.999)
13									0.001	(1.000)

c \ np_0	4.6		4.7		4.8		4.9		5.0	
0	0.010	(0.010)	0.009	(0.009)	0.008	(0.008)	0.008	(0.008)	0.007	(0.007)
1	0.046	(0.056)	0.043	(0.052)	0.039	(0.047)	0.037	(0.045)	0.034	(0.041)
2	0.106	(0.162)	0.101	(0.153)	0.095	(0.142)	0.090	(0.135)	0.084	(0.125)
3	0.163	(0.325)	0.157	(0.310)	0.152	(0.294)	0.146	(0.281)	0.140	(0.265)
4	0.188	(0.513)	0.185	(0.495)	0.182	(0.476)	0.179	(0.460)	0.176	(0.441)
5	0.172	(0.685)	0.174	(0.669)	0.175	(0.651)	0.175	(0.635)	0.176	(0.617)
6	0.132	(0.817)	0.136	(0.805)	0.140	(0.791)	0.143	(0.778)	0.146	(0.763)
7	0.087	(0.904)	0.091	(0.896)	0.096	(0.887)	0.100	(0.878)	0.105	(0.868)
8	0.050	(0.954)	0.054	(0.950)	0.058	(0.945)	0.061	(0.939)	0.065	(0.933)
9	0.026	(0.980)	0.028	(0.978)	0.031	(0.976)	0.034	(0.973)	0.036	(0.969)
10	0.012	(0.992)	0.013	(0.991)	0.015	(0.991)	0.016	(0.989)	0.018	(0.987)
11	0.005	(0.997)	0.006	(0.997)	0.006	(0.997)	0.007	(0.996)	0.008	(0.995)
12	0.002	(0.999)	0.002	(0.999)	0.002	(0.999)	0.003	(0.999)	0.003	(0.998)
13	0.001	(1.000)	0.001	(1.000)	0.001	(1.000)	0.001	(1.000)	0.001	(0.999)
14									0.001	(1.000)

c \ np_0	6.0		7.0		8.0		9.0		10.0	
0	0.002	(0.002)	0.001	(0.001)	0.000	(0.000)	0.000	(0.000)	0.000	(0.000)
1	0.015	(0.017)	0.006	(0.007)	0.003	(0.003)	0.001	(0.001)	0.000	(0.000)
2	0.045	(0.062)	0.022	(0.029)	0.011	(0.014)	0.005	(0.006)	0.002	(0.002)
3	0.089	(0.151)	0.052	(0.081)	0.029	(0.043)	0.015	(0.021)	0.007	(0.009)
4	0.134	(0.285)	0.091	(0.172)	0.057	(0.100)	0.034	(0.055)	0.019	(0.028)
5	0.161	(0.446)	0.128	(0.300)	0.092	(0.192)	0.061	(0.116)	0.038	(0.066)
6	0.161	(0.607)	0.149	(0.449)	0.122	(0.314)	0.091	(0.207)	0.063	(0.129)
7	0.138	(0.745)	0.149	(0.598)	0.140	(0.454)	0.117	(0.324)	0.090	(0.219)
8	0.103	(0.848)	0.131	(0.729)	0.140	(0.594)	0.132	(0.456)	0.113	(0.332)
9	0.069	(0.917)	0.102	(0.831)	0.124	(0.718)	0.132	(0.588)	0.125	(0.457)
10	0.041	(0.958)	0.071	(0.902)	0.099	(0.817)	0.119	(0.707)	0.125	(0.582)
11	0.023	(0.981)	0.045	(0.947)	0.072	(0.889)	0.097	(0.804)	0.114	(0.696)
12	0.011	(0.992)	0.026	(0.973)	0.048	(0.937)	0.073	(0.877)	0.095	(0.791)
13	0.005	(0.997)	0.014	(0.987)	0.030	(0.967)	0.050	(0.927)	0.073	(0.864)
14	0.002	(0.999)	0.007	(0.994)	0.017	(0.984)	0.032	(0.959)	0.052	(0.916)
15	0.001	(1.000)	0.003	(0.997)	0.009	(0.993)	0.019	(0.978)	0.035	(0.951)
16			0.002	(0.999)	0.004	(0.997)	0.011	(0.989)	0.022	(0.973)
17			0.001	(1.000)	0.002	(0.999)	0.006	(0.995)	0.013	(0.986)
18					0.001	(1.000)	0.003	(0.998)	0.007	(0.993)
19							0.001	(0.999)	0.004	(0.997)
20							0.001	(1.000)	0.002	(0.999)
21									0.001	(1.000)

APPENDIX B *(continued)*

c	np_0 11.0		12.0		13.0		14.0		15.0	
0	0.000	(0.000)	0.000	(0.000)	0.000	(0.000)	0.000	(0.000)	0.000	(0.000)
1	0.000	(0.000)	0.000	(0.000)	0.000	(0.000)	0.000	(0.000)	0.000	(0.000)
2	0.001	(0.001)	0.000	(0.000)	0.000	(0.000)	0.000	(0.000)	0.000	(0.000)
3	0.004	(0.005)	0.002	(0.002)	0.001	(0.001)	0.000	(0.000)	0.000	(0.000)
4	0.010	(0.015)	0.005	(0.007)	0.003	(0.004)	0.001	(0.001)	0.001	(0.001)
5	0.022	(0.037)	0.013	(0.020)	0.007	(0.011)	0.004	(0.005)	0.002	(0.003)
6	0.041	(0.078)	0.025	(0.045)	0.015	(0.026)	0.009	(0.014)	0.005	(0.008)
7	0.065	(0.143)	0.044	(0.089)	0.028	(0.054)	0.017	(0.031)	0.010	(0.018)
8	0.089	(0.232)	0.066	(0.155)	0.046	(0.100)	0.031	(0.062)	0.019	(0.037)
9	0.109	(0.341)	0.087	(0.242)	0.066	(0.166)	0.047	(0.109)	0.032	(0.069)
10	0.119	(0.460)	0.105	(0.347)	0.086	(0.252)	0.066	(0.175)	0.049	(0.118)
11	0.119	(0.579)	0.114	(0.461)	0.101	(0.353)	0.084	(0.259)	0.066	(0.184)
12	0.109	(0.688)	0.114	(0.575)	0.110	(0.463)	0.099	(0.358)	0.083	(0.267)
13	0.093	(0.781)	0.106	(0.681)	0.110	(0.573)	0.106	(0.464)	0.096	(0.363)
14	0.073	(0.854)	0.091	(0.772)	0.102	(0.675)	0.106	(0.570)	0.102	(0.465)
15	0.053	(0.907)	0.072	(0.844)	0.088	(0.763)	0.099	(0.669)	0.102	(0.567)
16	0.037	(0.944)	0.054	(0.898)	0.072	(0.835)	0.087	(0.756)	0.096	(0.663)
17	0.024	(0.968)	0.038	(0.936)	0.055	(0.890)	0.071	(0.827)	0.085	(0.748)
18	0.015	(0.983)	0.026	(0.962)	0.040	(0.930)	0.056	(0.883)	0.071	(0.819)
19	0.008	(0.991)	0.016	(0.978)	0.027	(0.957)	0.041	(0.924)	0.056	(0.875)
20	0.005	(0.996)	0.010	(0.988)	0.018	(0.975)	0.029	(0.953)	0.042	(0.917)
21	0.002	(0.998)	0.006	(0.994)	0.011	(0.986)	0.019	(0.972)	0.030	(0.947)
22	0.001	(0.999)	0.003	(0.997)	0.006	(0.992)	0.012	(0.984)	0.020	(0.967)
23	0.001	(1.000)	0.002	(0.999)	0.004	(0.996)	0.007	(0.991)	0.013	(0.980)
24			0.001	(1.000)	0.002	(0.998)	0.004	(0.995)	0.008	(0.988)
25					0.001	(0.999)	0.003	(0.998)	0.005	(0.993)
26					0.001	(1.000)	0.001	(0.999)	0.003	(0.996)
27							0.001	(1.000)	0.002	(0.998)
28									0.001	(0.999)
29									0.001	(1.000)

APPENDIX C.1 $\frac{1}{2}\%$ Interest Factors

	Single Payment		Equal Payment Series				Uniform gradient-series factor
	Compound-amount factor	Present-worth factor	Compound-amount factor	Sinking-fund factor	Present-worth factor	Capital-recovery factor	
n	To find F Given P F/P i,n	To find P Given F P/F i,n	To find F Given A F/A i,n	To find A Given F A/F i,n	To find P Given A P/A i,n	To find A Given P A/P i,n	To find A Given G A/G i,n
1	1.005	0.9950	1.000	1.0000	0.9950	1.0050	0.0000
2	1.010	0.9901	2.005	0.4988	1.9851	0.5038	0.4988
3	1.015	0.9852	3.015	0.3317	2.9703	0.3367	0.9967
4	1.020	0.9803	4.030	0.2481	3.9505	0.2531	1.4938
5	1.025	0.9754	5.050	0.1980	4.9259	0.2030	1.9900
6	1.030	0.9705	6.076	0.1646	5.8964	0.1696	2.4855
7	1.036	0.9657	7.106	0.1407	6.8621	0.1457	2.9801
8	1.041	0.9609	8.141	0.1228	7.8230	0.1278	3.4738
9	1.046	0.9561	9.182	0.1089	8.7791	0.1139	3.9668
10	1.051	0.9514	10.228	0.0978	9.7304	0.1028	4.4589
11	1.056	0.9466	11.279	0.0887	10.6770	0.0937	4.9501
12	1.062	0.9419	12.336	0.0811	11.6189	0.0861	5.4406
13	1.067	0.9372	13.397	0.0747	12.5562	0.0797	5.9302
14	1.072	0.9326	14.464	0.0691	13.4887	0.0741	6.4190
15	1.078	0.9279	15.537	0.0644	14.4166	0.0694	6.9069
16	1.083	0.9233	16.614	0.0602	15.3399	0.0652	7.3940
17	1.088	0.9187	17.697	0.0565	16.2586	0.0615	7.8803
18	1.094	0.9141	18.786	0.0532	17.1728	0.0582	8.3658
19	1.099	0.9096	19.880	0.0503	18.0824	0.0553	8.8504
20	1.105	0.9051	20.979	0.0477	18.9874	0.0527	9.3342
21	1.110	0.9006	22.084	0.0453	19.8880	0.0503	9.8172
22	1.116	0.8961	23.194	0.0431	20.7841	0.0481	10.2993
23	1.122	0.8916	24.310	0.0411	21.6757	0.0461	10.7806
24	1.127	0.8872	25.432	0.0393	22.5629	0.0443	11.2611
25	1.133	0.8828	26.559	0.0377	23.4456	0.0427	11.7407
26	1.138	0.8784	27.692	0.0361	24.3240	0.0411	12.2195
27	1.144	0.8740	28.830	0.0347	25.1980	0.0397	12.6975
28	1.150	0.8697	29.975	0.0334	26.0677	0.0384	13.1747
29	1.156	0.8653	31.124	0.0321	26.9330	0.0371	13.6510
30	1.161	0.8610	32.280	0.0310	27.7941	0.0360	14.1265
31	1.167	0.8568	33.441	0.0299	28.6508	0.0349	14.6012
32	1.173	0.8525	34.609	0.0289	29.5033	0.0339	15.0750
33	1.179	0.8483	35.782	0.0280	30.3515	0.0330	15.5480
34	1.185	0.8440	36.961	0.0271	31.1956	0.0321	16.0202
35	1.191	0.8398	38.145	0.0262	32.0354	0.0312	16.4915
40	1.221	0.8191	44.159	0.0227	36.1722	0.0277	18.8358
45	1.252	0.7990	50.324	0.0199	40.2072	0.0249	21.1595
50	1.283	0.7793	56.645	0.0177	44.1428	0.0227	23.4624
55	1.316	0.7601	63.126	0.0159	47.9815	0.0209	25.7447
60	1.349	0.7414	69.770	0.0143	51.7256	0.0193	28.0064
65	1.383	0.7231	76.582	0.0131	55.3775	0.0181	30.2475
70	1.418	0.7053	83.566	0.0120	58.9394	0.0170	32.4680
75	1.454	0.6879	90.727	0.0110	62.4137	0.0160	34.6679
80	1.490	0.6710	98.068	0.0102	65.8023	0.0152	36.8474
85	1.528	0.6545	105.594	0.0095	69.1075	0.0145	39.0065
90	1.567	0.6384	113.311	0.0088	72.3313	0.0138	41.1451
95	1.606	0.6226	121.222	0.0083	75.4757	0.0133	43.2633
100	1.647	0.6073	129.334	0.0077	78.5427	0.0127	45.3613

Source: G. J. Thuesen & W. J. Fabrycky, *Engineering Economy*, 7e, © 1989, pp. 645–662. Reprinted by permission of Prentice-Hall, Inc., Englewood Cliffs, New Jersey.

268

APPENDIX C.2 $\frac{3}{4}\%$ Interest Factors

	Single Payment		Equal Payment Series				Uniform gradient-series factor
	Compound-amount factor	Present-worth factor	Compound-amount factor	Sinking-fund factor	Present-worth factor	Capital-recovery factor	
n	To find F Given P F/P i,n	To find P Given F P/F i,n	To find F Given A F/A i,n	To find A Given F A/F i,n	To find P Given A P/A i,n	To find A Given P A/P i,n	To find A Given G A/G i,n
1	1.008	0.9926	1.000	1.0000	0.9926	1.0075	0.0000
2	1.015	0.9852	2.008	0.4981	1.9777	0.5056	0.4981
3	1.023	0.9778	3.023	0.3309	2.9556	0.3384	0.9950
4	1.030	0.9706	4.045	0.2472	3.9261	0.2547	1.4907
5	1.038	0.9633	5.076	0.1970	4.8894	0.2045	1.9851
6	1.046	0.9562	6.114	0.1636	5.8456	0.1711	2.4782
7	1.054	0.9491	7.159	0.1397	6.7946	0.1472	2.9701
8	1.062	0.9420	8.213	0.1218	7.7366	0.1293	3.4608
9	1.070	0.9350	9.275	0.1078	8.6716	0.1153	3.9502
10	1.078	0.9280	10.344	0.0967	9.5996	0.1042	4.4384
11	1.086	0.9211	11.422	0.0876	10.5207	0.0951	4.9253
12	1.094	0.9142	12.508	0.0800	11.4349	0.0875	5.4110
13	1.102	0.9074	13.601	0.0735	12.3424	0.0810	5.8954
14	1.110	0.9007	14.703	0.0680	13.2430	0.0755	6.3786
15	1.119	0.8940	15.814	0.0632	14.1370	0.0707	6.8606
16	1.127	0.8873	16.932	0.0591	15.0243	0.0666	7.3413
17	1.135	0.8807	18.059	0.0554	15.9050	0.0629	7.8207
18	1.144	0.8742	19.195	0.0521	16.7792	0.0596	8.2989
19	1.153	0.8677	20.339	0.0492	17.6468	0.0567	8.7759
20	1.161	0.8612	21.491	0.0465	18.5080	0.0540	9.2517
21	1.170	0.8548	22.652	0.0442	19.3628	0.0517	9.7261
22	1.179	0.8484	23.822	0.0420	20.2112	0.0495	10.1994
23	1.188	0.8421	25.001	0.0400	21.0533	0.0475	10.6714
24	1.196	0.8358	26.188	0.0382	21.8892	0.0457	11.1422
25	1.205	0.8296	27.385	0.0365	22.7188	0.0440	11.6117
26	1.214	0.8234	28.590	0.0350	23.5422	0.0425	12.0800
27	1.224	0.8173	29.805	0.0336	24.3595	0.0411	12.5470
28	1.233	0.8112	31.028	0.0322	25.1707	0.0397	13.0128
29	1.242	0.8052	32.261	0.0310	25.9759	0.0385	13.4774
30	1.251	0.7992	33.503	0.0299	26.7751	0.0374	13.9407
31	1.261	0.7932	34.754	0.0288	27.5683	0.0363	14.4028
32	1.270	0.7873	36.015	0.0278	28.3557	0.0353	14.8636
33	1.280	0.7815	37.285	0.0268	29.1371	0.0343	15.3232
34	1.289	0.7757	38.565	0.0259	29.9128	0.0334	15.7816
35	1.299	0.7699	39.854	0.0251	30.6827	0.0326	16.2387
40	1.348	0.7417	46.446	0.0215	34.4469	0.0290	18.5058
45	1.400	0.7145	53.290	0.0188	38.0732	0.0263	20.7421
50	1.453	0.6883	60.394	0.0166	41.5665	0.0241	22.9476
55	1.508	0.6630	67.769	0.0148	44.9316	0.0223	25.1223
60	1.566	0.6387	75.424	0.0133	48.1734	0.0208	27.2665
65	1.625	0.6153	83.371	0.0120	51.2963	0.0195	29.3801
70	1.687	0.5927	91.620	0.0109	54.3046	0.0184	31.4634
75	1.751	0.5710	100.183	0.0100	57.2027	0.0175	33.5163
80	1.818	0.5501	109.073	0.0092	59.9945	0.0167	35.5391
85	1.887	0.5299	118.300	0.0085	62.6838	0.0160	37.5318
90	1.959	0.5105	127.879	0.0078	65.2746	0.0153	39.4946
95	2.034	0.4917	137.823	0.0073	67.7704	0.0148	41.4277
100	2.111	0.4737	148.145	0.0068	70.1746	0.0143	43.3311

APPENDIX C.3   1% Interest Factors

	Single Payment		Equal Payment Series				Uniform gradient-series factor
	Compound-amount factor	Present-worth factor	Compound-amount factor	Sinking-fund factor	Present-worth factor	Capital-recovery factor	
n	To find F Given P F/P i, n	To find P Given F P/F i, n	To find F Given A F/A i, n	To find A Given F A/F i, n	To find P Given A P/A i, n	To find A Given P A/P i, n	To find A Given G A/G i, n
1	1.010	0.9901	1.000	1.0000	0.9901	1.0100	0.0000
2	1.020	0.9803	2.010	0.4975	1.9704	0.5075	0.4975
3	1.030	0.9706	3.030	0.3300	2.9410	0.3400	0.9934
4	1.041	0.9610	4.060	0.2463	3.9020	0.2563	1.4876
5	1.051	0.9515	5.101	0.1960	4.8534	0.2060	1.9801
6	1.062	0.9421	6.152	0.1626	5.7955	0.1726	2.4710
7	1.072	0.9327	7.214	0.1386	6.7282	0.1486	2.9602
8	1.083	0.9235	8.286	0.1207	7.6517	0.1307	3.4478
9	1.094	0.9143	9.369	0.1068	8.5660	0.1168	3.9337
10	1.105	0.9053	10.462	0.0956	9.4713	0.1056	4.4179
11	1.116	0.8963	11.567	0.0865	10.3676	0.0965	4.9005
12	1.127	0.8875	12.683	0.0789	11.2551	0.0889	5.3815
13	1.138	0.8787	13.809	0.0724	12.1338	0.0824	5.8607
14	1.149	0.8700	14.947	0.0669	13.0037	0.0769	6.3384
15	1.161	0.8614	16.097	0.0621	13.8651	0.0721	6.8143
16	1.173	0.8528	17.258	0.0580	14.7179	0.0680	7.2887
17	1.184	0.8444	18.430	0.0543	15.5623	0.0643	7.7613
18	1.196	0.8360	19.615	0.0510	16.3983	0.0610	8.2323
19	1.208	0.8277	20.811	0.0481	17.2260	0.0581	8.7017
20	1.220	0.8196	22.019	0.0454	18.0456	0.0554	9.1694
21	1.232	0.8114	23.239	0.0430	18.8570	0.0530	9.6354
22	1.245	0.8034	24.472	0.0409	19.6604	0.0509	10.0998
23	1.257	0.7955	25.716	0.0389	20.4558	0.0489	10.5626
24	1.270	0.7876	26.973	0.0371	21.2434	0.0471	11.0237
25	1.282	0.7798	28.243	0.0354	22.0232	0.0454	11.4831
26	1.295	0.7721	29.526	0.0339	22.7952	0.0439	11.9409
27	1.308	0.7644	30.821	0.0325	23.5596	0.0425	12.3971
28	1.321	0.7568	32.129	0.0311	24.3165	0.0411	12.8516
29	1.335	0.7494	33.450	0.0299	25.0658	0.0399	13.3045
30	1.348	0.7419	34.785	0.0288	25.8077	0.0388	13.7557
31	1.361	0.7346	36.133	0.0277	26.5423	0.0377	14.2052
32	1.375	0.7273	37.494	0.0267	27.2696	0.0367	14.6532
33	1.389	0.7201	38.869	0.0257	27.9897	0.0357	15.0995
34	1.403	0.7130	40.258	0.0248	28.7027	0.0348	15.5441
35	1.417	0.7059	41.660	0.0240	29.4086	0.0340	15.9871
40	1.489	0.6717	48.886	0.0205	32.8347	0.0305	18.1776
45	1.565	0.6391	56.481	0.0177	36.0945	0.0277	20.3273
50	1.645	0.6080	64.463	0.0155	39.1961	0.0255	22.4363
55	1.729	0.5785	72.852	0.0137	42.1472	0.0237	24.5049
60	1.817	0.5505	81.670	0.0123	44.9550	0.0223	26.5333
65	1.909	0.5237	90.937	0.0110	47.6266	0.0210	28.5217
70	2.007	0.4983	100.676	0.0099	50.1685	0.0199	30.4703
75	2.109	0.4741	110.913	0.0090	52.5871	0.0190	32.3793
80	2.217	0.4511	121.672	0.0082	54.8882	0.0182	34.2492
85	2.330	0.4292	132.979	0.0075	57.0777	0.0175	36.0801
90	2.449	0.4084	144.863	0.0069	59.1609	0.0169	37.8725
95	2.574	0.3886	157.354	0.0064	61.1430	0.0164	39.6265
100	2.705	0.3697	170.481	0.0059	63.0289	0.0159	41.3426

APPENDIX C.4 $1\frac{1}{4}\%$ Interest Factors

	Single Payment		Equal Payment Series				Uniform gradient-series factor
	Compound-amount factor	Present-worth factor	Compound-amount factor	Sinking-fund factor	Present-worth factor	Capital-recovery factor	
n	To find F Given P $F/P\ i,n$	To find P Given F $P/F\ i,n$	To find F Given A $F/A\ i,n$	To find A Given F $A/F\ i,n$	To find P Given A $P/A\ i,n$	To find A Given P $A/P\ i,n$	To find A Given G $A/G\ i,n$
1	1.013	0.9877	1.000	1.0001	0.9877	1.0126	0.0000
2	1.025	0.9755	2.013	0.4970	1.9631	0.5095	0.4932
3	1.038	0.9635	3.038	0.3293	2.9265	0.3418	0.9895
4	1.051	0.9516	4.076	0.2454	3.8780	0.2579	1.4830
5	1.064	0.9398	5.127	0.1951	4.8177	0.2076	1.9729
6	1.077	0.9282	6.191	0.1616	5.7459	0.1741	2.4618
7	1.091	0.9168	7.268	0.1376	6.6627	0.1501	2.9491
8	1.105	0.9055	8.359	0.1197	7.5680	0.1322	3.4330
9	1.118	0.8943	9.463	0.1057	8.4623	0.1182	3.9158
10	1.132	0.8832	10.582	0.0946	9.3454	0.1071	4.3960
11	1.147	0.8723	11.714	0.0854	10.2177	0.0979	4.8744
12	1.161	0.8616	12.860	0.0778	11.0792	0.0903	5.3506
13	1.175	0.8509	14.021	0.0714	11.9300	0.0839	5.8248
14	1.190	0.8404	15.196	0.0659	12.7704	0.0784	6.2968
15	1.205	0.8300	16.386	0.0611	13.6004	0.0736	6.7669
16	1.220	0.8198	17.591	0.0569	14.4201	0.0694	7.2350
17	1.235	0.8097	18.811	0.0532	15.2298	0.0657	7.7009
18	1.251	0.7997	20.046	0.0499	16.0293	0.0624	8.1645
19	1.266	0.7898	21.296	0.0470	16.8191	0.0595	8.6264
20	1.282	0.7801	22.563	0.0444	17.5991	0.0569	9.0861
21	1.298	0.7704	23.845	0.0420	18.3695	0.0545	9.5439
22	1.314	0.7609	25.143	0.0398	19.1303	0.0523	9.9993
23	1.331	0.7515	26.457	0.0378	19.8818	0.0503	10.4528
24	1.347	0.7423	27.788	0.0360	20.6240	0.0485	10.9044
25	1.364	0.7331	29.135	0.0344	21.3570	0.0469	11.3539
26	1.381	0.7240	30.499	0.0328	22.0810	0.0453	11.8012
27	1.399	0.7151	31.880	0.0314	22.7960	0.0439	12.2465
28	1.416	0.7063	33.279	0.0301	23.5022	0.0426	12.6898
29	1.434	0.6976	34.695	0.0289	24.1998	0.0414	13.1311
30	1.452	0.6889	36.128	0.0277	24.8886	0.0402	13.5703
31	1.470	0.6804	37.580	0.0267	25.5690	0.0392	14.0074
32	1.488	0.6720	39.050	0.0257	26.2410	0.0382	14.4425
33	1.507	0.6637	40.538	0.0247	26.9047	0.0372	14.8756
34	1.526	0.6555	42.045	0.0238	27.5601	0.0363	15.3066
35	1.545	0.6475	43.570	0.0230	28.2075	0.0355	15.7357
40	1.644	0.6085	51.489	0.0195	31.3266	0.0320	17.8503
45	1.749	0.5718	59.915	0.0167	34.2578	0.0292	19.9144
50	1.861	0.5374	68.880	0.0146	37.0125	0.0271	21.9284
55	1.980	0.5050	78.421	0.0128	39.6013	0.0253	23.8925
60	2.107	0.4746	88.573	0.0113	42.0342	0.0238	25.8072
65	2.242	0.4460	99.375	0.0101	44.3206	0.0226	27.6730
70	2.386	0.4192	110.870	0.0091	46.4693	0.0216	29.4902
75	2.539	0.3939	123.101	0.0082	48.4886	0.0207	31.2594
80	2.702	0.3702	136.116	0.0074	50.3862	0.0199	32.9812
85	2.875	0.3479	149.965	0.0067	52.1696	0.0192	34.6560
90	3.059	0.3270	164.701	0.0061	53.8456	0.0186	36.2844
95	3.255	0.3073	180.382	0.0056	55.4207	0.0181	37.8671
100	3.463	0.2888	197.067	0.0051	56.9009	0.0176	39.4048

APPENDIX C.5 $1\frac{1}{2}\%$ Interest Factors

	Single Payment		Equal Payment Series				Uniform gradient-series factor
	Compound-amount factor	Present-worth factor	Compound-amount factor	Sinking-fund factor	Present-worth factor	Capital-recovery factor	
n	To find F Given P F/P i, n	To find P Given F P/F i, n	To find F Given A F/A i, n	To find A Given F A/F i, n	To find P Given A P/A i, n	To find A Given P A/P i, n	To find A Given G A/G i, n
1	1.015	0.9852	1.000	1.0000	0.9852	1.0150	0.0000
2	1.030	0.9707	2.015	0.4963	1.9559	0.5113	0.4963
3	1.046	0.9563	3.045	0.3284	2.9122	0.3434	0.9901
4	1.061	0.9422	4.091	0.2445	3.8544	0.2595	1.4814
5	1.077	0.9283	5.152	0.1941	4.7827	0.2091	1.9702
6	1.093	0.9146	6.230	0.1605	5.6972	0.1755	2.4566
7	1.110	0.9010	7.323	0.1366	6.5982	0.1516	2.9405
8	1.127	0.8877	8.433	0.1186	7.4859	0.1336	3.4219
9	1.143	0.8746	9.559	0.1046	8.3605	0.1196	3.9008
10	1.161	0.8617	10.703	0.0934	9.2222	0.1084	4.3772
11	1.178	0.8489	11.863	0.0843	10.0711	0.0993	4.8512
12	1.196	0.8364	13.041	0.0767	10.9075	0.0917	5.3227
13	1.214	0.8240	14.237	0.0703	11.7315	0.0853	5.7917
14	1.232	0.8119	15.450	0.0647	12.5434	0.0797	6.2582
15	1.250	0.7999	16.682	0.0600	13.3432	0.0750	6.7223
16	1.269	0.7880	17.932	0.0558	14.1313	0.0708	7.1839
17	1.288	0.7764	19.201	0.0521	14.9077	0.0671	7.6431
18	1.307	0.7649	20.489	0.0488	15.6726	0.0638	8.0997
19	1.327	0.7536	21.797	0.0459	16.4262	0.0609	8.5539
20	1.347	0.7425	23.124	0.0433	17.1686	0.0583	9.0057
21	1.367	0.7315	24.471	0.0409	17.9001	0.0559	9.4550
22	1.388	0.7207	25.838	0.0387	18.6208	0.0537	9.9018
23	1.408	0.7100	27.225	0.0367	19.3309	0.0517	10.3462
24	1.430	0.6996	28.634	0.0349	20.0304	0.0499	10.7881
25	1.451	0.6892	30.063	0.0333	20.7196	0.0483	11.2276
26	1.473	0.6790	31.514	0.0317	21.3986	0.0467	11.6646
27	1.495	0.6690	32.987	0.0303	22.0676	0.0453	12.0992
28	1.517	0.6591	34.481	0.0290	22.7267	0.0440	12.5313
29	1.540	0.6494	35.999	0.0278	23.3761	0.0428	12.9610
30	1.563	0.6398	37.539	0.0266	24.0158	0.0416	13.3883
31	1.587	0.6303	39.102	0.0256	24.6462	0.0406	13.8131
32	1.610	0.6210	40.688	0.0246	25.2671	0.0396	14.2355
33	1.634	0.6118	42.299	0.0237	25.8790	0.0387	14.6555
34	1.659	0.6028	43.933	0.0228	26.4817	0.0378	15.0731
35	1.684	0.5939	45.592	0.0219	27.0756	0.0369	15.4882
40	1.814	0.5513	54.268	0.0184	29.9159	0.0334	17.5277
45	1.954	0.5117	63.614	0.0157	32.5523	0.0307	19.5074
50	2.105	0.4750	73.683	0.0136	34.9997	0.0286	21.4277
55	2.268	0.4409	84.530	0.0118	37.2715	0.0268	23.2894
60	2.443	0.4093	96.215	0.0104	39.3803	0.0254	25.0930
65	2.632	0.3799	108.803	0.0092	41.3378	0.0242	26.8392
70	2.835	0.3527	122.364	0.0082	43.1549	0.0232	28.5290
75	3.055	0.3274	136.973	0.0073	44.8416	0.0223	30.1631
80	3.291	0.3039	152.711	0.0066	46.4073	0.0216	31.7423
85	3.545	0.2821	169.665	0.0059	47.8607	0.0209	33.2676
90	3.819	0.2619	187.930	0.0053	49.2099	0.0203	34.7399
95	4.114	0.2431	207.606	0.0048	50.4622	0.0198	36.1602
100	4.432	0.2256	228.803	0.0044	51.6247	0.0194	37.5295

APPENDIX C.6 2% Interest Factors

	Single Payment		Equal Payment Series				Uniform gradient-series factor
	Compound-amount factor	Present-worth factor	Compound-amount factor	Sinking-fund factor	Present-worth factor	Capital-recovery factor	
n	To find F Given P $F/P \ i, n$	To find P Given F $P/F \ i, n$	To find F Given A $F/A \ i, n$	To find A Given F $A/F \ i, n$	To find P Given A $P/A \ i, n$	To find A Given P $A/P \ i, n$	To find A Given G $A/G \ i, n$
1	1.020	0.9804	1.000	1.0000	0.9804	1.0200	0.0000
2	1.040	0.9612	2.020	0.4951	1.9416	0.5151	0.4951
3	1.061	0.9423	3.060	0.3268	2.8839	0.3468	0.9868
4	1.082	0.9239	4.122	0.2426	3.8077	0.2626	1.4753
5	1.104	0.9057	5.204	0.1922	4.7135	0.2122	1.9604
6	1.126	0.8880	6.308	0.1585	5.6014	0.1785	2.4423
7	1.149	0.8706	7.434	0.1345	6.4720	0.1545	2.9208
8	1.172	0.8535	8.583	0.1165	7.3255	0.1365	3.3961
9	1.195	0.8368	9.755	0.1025	8.1622	0.1225	3.8681
10	1.219	0.8204	10.950	0.0913	8.9826	0.1113	4.3367
11	1.243	0.8043	12.169	0.0822	9.7869	0.1022	4.8021
12	1.268	0.7885	13.412	0.0746	10.5754	0.0946	5.2643
13	1.294	0.7730	14.680	0.0681	11.3484	0.0881	5.7231
14	1.319	0.7579	15.974	0.0626	12.1063	0.0826	6.1786
15	1.346	0.7430	17.293	0.0578	12.8493	0.0778	6.6309
16	1.373	0.7285	18.639	0.0537	13.5777	0.0737	7.0799
17	1.400	0.7142	20.012	0.0500	14.2919	0.0700	7.5256
18	1.428	0.7002	21.412	0.0467	14.9920	0.0667	7.9681
19	1.457	0.6864	22.841	0.0438	15.6785	0.0638	8.4073
20	1.486	0.6730	24.297	0.0412	16.3514	0.0612	8.8433
21	1.516	0.6598	25.783	0.0388	17.0112	0.0588	9.2760
22	1.546	0.6468	27.299	0.0366	17.6581	0.0566	9.7055
23	1.577	0.6342	28.845	0.0347	18.2922	0.0547	10.1317
24	1.608	0.6217	30.422	0.0329	18.9139	0.0529	10.5547
25	1.641	0.6095	32.030	0.0312	19.5235	0.0512	10.9745
26	1.673	0.5976	33.671	0.0297	20.1210	0.0497	11.3910
27	1.707	0.5859	35.344	0.0283	20.7069	0.0483	11.8043
28	1.741	0.5744	37.051	0.0270	21.2813	0.0470	12.2145
29	1.776	0.5631	38.792	0.0258	21.8444	0.0458	12.6214
30	1.811	0.5521	40.568	0.0247	22.3965	0.0447	13.0251
31	1.848	0.5413	42.379	0.0236	22.9377	0.0436	13.4257
32	1.885	0.5306	44.227	0.0226	23.4683	0.0426	13.8230
33	1.922	0.5202	46.112	0.0217	23.9886	0.0417	14.2172
34	1.961	0.5100	48.034	0.0208	24.4986	0.0408	14.6083
35	2.000	0.5000	49.994	0.0200	24.9986	0.0400	14.9961
40	2.208	0.4529	60.402	0.0166	27.3555	0.0366	16.8885
45	2.438	0.4102	71.893	0.0139	29.4902	0.0339	18.7034
50	2.692	0.3715	84.579	0.0118	31.4236	0.0318	20.4420
55	2.972	0.3365	98.587	0.0102	33.1748	0.0302	22.1057
60	3.281	0.3048	114.052	0.0088	34.7609	0.0288	23.6961
65	3.623	0.2761	131.126	0.0076	36.1975	0.0276	25.2147
70	4.000	0.2500	149.978	0.0067	37.4986	0.0267	26.6632
75	4.416	0.2265	170.792	0.0059	38.6771	0.0259	28.0434
80	4.875	0.2051	193.772	0.0052	39.7445	0.0252	29.3572
85	5.383	0.1858	219.144	0.0046	40.7113	0.0246	30.6064
90	5.943	0.1683	247.157	0.0041	41.5869	0.0241	31.7929
95	6.562	0.1524	278.085	0.0036	42.3800	0.0236	32.9189
100	7.245	0.1380	312.232	0.0032	43.0984	0.0232	33.9863

APPENDIX C.7   3% Interest Factors

	Single Payment		Equal Payment Series				Uniform gradient-series factor
	Compound-amount factor	Present-worth factor	Compound-amount factor	Sinking-fund factor	Present-worth factor	Capital-recovery factor	
n	To find F Given P F/P i,n	To find P Given F P/F i,n	To find F Given A F/A i,n	To find A Given F A/F i,n	To find P Given A P/A i,n	To find A Given P A/P i,n	To find A Given G A/G i,n
1	1.030	0.9709	1.000	1.0000	0.9709	1.0300	0.0000
2	1.061	0.9426	2.030	0.4926	1.9135	0.5226	0.4926
3	1.093	0.9152	3.091	0.3235	2.8286	0.3535	0.9803
4	1.126	0.8885	4.184	0.2390	3.7171	0.2690	1.4631
5	1.159	0.8626	5.309	0.1884	4.5797	0.2184	1.9409
6	1.194	0.8375	6.468	0.1546	5.4172	0.1846	2.4138
7	1.230	0.8131	7.662	0.1305	6.2303	0.1605	2.8819
8	1.267	0.7894	8.892	0.1125	7.0197	0.1425	3.3450
9	1.305	0.7664	10.159	0.0984	7.7861	0.1284	3.8032
10	1.344	0.7441	11.464	0.0872	8.5302	0.1172	4.2565
11	1.384	0.7224	12.808	0.0781	9.2526	0.1081	4.7049
12	1.426	0.7014	14.192	0.0705	9.9540	0.1005	5.1485
13	1.469	0.6810	15.618	0.0640	10.6350	0.0940	5.5872
14	1.513	0.6611	17.086	0.0585	11.2961	0.0885	6.0211
15	1.558	0.6419	18.599	0.0538	11.9379	0.0838	6.4501
16	1.605	0.6232	20.157	0.0496	12.5611	0.0796	6.8742
17	1.653	0.6050	21.762	0.0460	13.1661	0.0760	7.2936
18	1.702	0.5874	23.414	0.0427	13.7535	0.0727	7.7081
19	1.754	0.5703	25.117	0.0398	14.3238	0.0698	8.1179
20	1.806	0.5537	26.870	0.0372	14.8775	0.0672	8.5229
21	1.860	0.5376	28.676	0.0349	15.4150	0.0649	8.9231
22	1.916	0.5219	30.537	0.0328	15.9369	0.0628	9.3186
23	1.974	0.5067	32.453	0.0308	16.4436	0.0608	9.7094
24	2.033	0.4919	34.426	0.0291	16.9356	0.0591	10.0954
25	2.094	0.4776	36.459	0.0274	17.4132	0.0574	10.4768
26	2.157	0.4637	38.553	0.0259	17.8769	0.0559	10.8535
27	2.221	0.4502	40.710	0.0246	18.3270	0.0546	11.2256
28	2.288	0.4371	42.931	0.0233	18.7641	0.0533	11.5930
29	2.357	0.4244	45.219	0.0221	19.1885	0.0521	11.9558
30	2.427	0.4120	47.575	0.0210	19.6005	0.0510	12.3141
31	2.500	0.4000	50.003	0.0200	20.0004	0.0500	12.6678
32	2.575	0.3883	52.503	0.0191	20.3888	0.0491	13.0169
33	2.652	0.3770	55.078	0.0182	20.7658	0.0482	13.3616
34	2.732	0.3661	57.730	0.0173	21.1318	0.0473	13.7018
35	2.814	0.3554	60.462	0.0165	21.4872	0.0465	14.0375
40	3.262	0.3066	75.401	0.0133	23.1148	0.0433	15.6502
45	3.782	0.2644	92.720	0.0108	24.5187	0.0408	17.1556
50	4.384	0.2281	112.797	0.0089	25.7298	0.0389	18.5575
55	5.082	0.1968	136.072	0.0074	26.7744	0.0374	19.8600
60	5.892	0.1697	163.053	0.0061	27.6756	0.0361	21.0674
65	6.830	0.1464	194.333	0.0052	28.4529	0.0352	22.1841
70	7.918	0.1263	230.594	0.0043	29.1234	0.0343	23.2145
75	9.179	0.1090	272.631	0.0037	29.7018	0.0337	24.1634
80	10.641	0.0940	321.363	0.0031	30.2008	0.0331	25.0354
85	12.336	0.0811	377.857	0.0027	30.6312	0.0327	25.8349
90	14.300	0.0699	443.349	0.0023	31.0024	0.0323	26.5667
95	16.578	0.0603	519.272	0.0019	31.3227	0.0319	27.2351
100	19.219	0.0520	607.288	0.0017	31.5989	0.0317	27.8445

APPENDIX C.8 4% Interest Factors

	Single Payment		Equal Payment Series				Uniform gradient-series factor
	Compound-amount factor	Present-worth factor	Compound-amount factor	Sinking-fund factor	Present-worth factor	Capital-recovery factor	
n	To find F Given P F/P i, n	To find P Given F P/F i, n	To find F Given A F/A i, n	To find A Given F A/F i, n	To find P Given A P/A i, n	To find A Given P A/P i, n	To find A Given G A/G i, n
1	1.040	0.9615	1.000	1.0000	0.9615	1.0400	0.0000
2	1.082	0.9246	2.040	0.4902	1.8861	0.5302	0.4902
3	1.125	0.8890	3.122	0.3204	2.7751	0.3604	0.9739
4	1.170	0.8548	4.246	0.2355	3.6299	0.2755	1.4510
5	1.217	0.8219	5.416	0.1846	4.4518	0.2246	1.9216
6	1.265	0.7903	6.633	0.1508	5.2421	0.1908	2.3857
7	1.316	0.7599	7.898	0.1266	6.0021	0.1666	2.8433
8	1.369	0.7307	9.214	0.1085	6.7328	0.1485	3.2944
9	1.423	0.7026	10.583	0.0945	7.4353	0.1345	3.7391
10	1.480	0.6756	12.006	0.0833	8.1109	0.1233	4.1773
11	1.539	0.6496	13.486	0.0742	8.7605	0.1142	4.6090
12	1.601	0.6246	15.026	0.0666	9.3851	0.1066	5.0344
13	1.665	0.6006	16.627	0.0602	9.9857	0.1002	5.4533
14	1.732	0.5775	18.292	0.0547	10.5631	0.0947	5.8659
15	1.801	0.5553	20.024	0.0500	11.1184	0.0900	6.2721
16	1.873	0.5339	21.825	0.0458	11.6523	0.0858	6.6720
17	1.948	0.5134	23.698	0.0422	12.1657	0.0822	7.0656
18	2.026	0.4936	25.645	0.0390	12.6593	0.0790	7.4530
19	2.107	0.4747	27.671	0.0361	13.1339	0.0761	7.8342
20	2.191	0.4564	29.778	0.0336	13.5903	0.0736	8.2091
21	2.279	0.4388	31.969	0.0313	14.0292	0.0713	8.5780
22	2.370	0.4220	34.248	0.0292	14.4511	0.0692	8.9407
23	2.465	0.4057	36.618	0.0273	14.8569	0.0673	9.2973
24	2.563	0.3901	39.083	0.0256	15.2470	0.0656	9.6479
25	2.666	0.3751	41.646	0.0240	15.6221	0.0640	9.9925
26	2.772	0.3607	44.312	0.0226	15.9828	0.0626	10.3312
27	2.883	0.3468	47.084	0.0212	16.3296	0.0612	10.6640
28	2.999	0.3335	49.968	0.0200	16.6631	0.0600	10.9909
29	3.119	0.3207	52.966	0.0189	16.9837	0.0589	11.3121
30	3.243	0.3083	56.085	0.0178	17.2920	0.0578	11.6274
31	3.373	0.2965	59.328	0.0169	17.5885	0.0569	11.9371
32	3.508	0.2851	62.701	0.0160	17.8736	0.0560	12.2411
33	3.648	0.2741	66.210	0.0151	18.1477	0.0551	12.5396
34	3.794	0.2636	69.858	0.0143	18.4112	0.0543	12.8325
35	3.946	0.2534	73.652	0.0136	18.6646	0.0536	13.1199
40	4.801	0.2083	95.026	0.0105	19.7928	0.0505	14.4765
45	5.841	0.1712	121.029	0.0083	20.7200	0.0483	15.7047
50	7.107	0.1407	152.667	0.0066	21.4822	0.0466	16.8123
55	8.646	0.1157	191.159	0.0052	22.1086	0.0452	17.8070
60	10.520	0.0951	237.991	0.0042	22.6235	0.0442	18.6972
65	12.799	0.0781	294.968	0.0034	23.0467	0.0434	19.4909
70	15.572	0.0642	364.290	0.0028	23.3945	0.0428	20.1961
75	18.945	0.0528	448.631	0.0022	23.6804	0.0422	20.8206
80	23.050	0.0434	551.245	0.0018	23.9154	0.0418	21.3719
85	28.044	0.0357	676.090	0.0015	24.1085	0.0415	21.8569
90	34.119	0.0293	817.983	0.0012	24.2673	0.0412	22.2826
95	41.511	0.0241	1012.785	0.0010	24.3978	0.0410	22.6550
100	50.505	0.0198	1237.624	0.0008	24.5050	0.0408	22.9800

APPENDIX C.9 5% Interest Factors

	Single Payment		Equal Payment Series				Uniform gradient-series factor
	Compound-amount factor	Present-worth factor	Compound-amount factor	Sinking-fund factor	Present-worth factor	Capital-recovery factor	
n	To find *F* Given *P* F/P *i, n*	To find *P* Given *F* P/F *i, n*	To find *F* Given *A* F/A *i, n*	To find *A* Given *F* A/F *i, n*	To find *P* Given *A* P/A *i, n*	To find *A* Given *P* A/P *i, n*	To find *A* Given *G* A/G *i, n*
1	1.050	0.9524	1.000	1.0000	0.9524	1.0500	0.0000
2	1.103	0.9070	2.050	0.4878	1.8594	0.5378	0.4878
3	1.158	0.8638	3.153	0.3172	2.7233	0.3672	0.9675
4	1.216	0.8227	4.310	0.2320	3.5460	0.2820	1.4391
5	1.276	0.7835	5.526	0.1810	4.3295	0.2310	1.9025
6	1.340	0.7462	6.802	0.1470	5.0757	0.1970	2.3579
7	1.407	0.7107	8.142	0.1228	5.7864	0.1728	2.8052
8	1.477	0.6768	9.549	0.1047	6.4632	0.1547	3.2445
9	1.551	0.6446	11.027	0.0907	7.1078	0.1407	3.6758
10	1.629	0.6139	12.587	0.0795	7.7217	0.1295	4.0991
11	1.710	0.5847	14.207	0.0704	8.3064	0.1204	4.5145
12	1.796	0.5568	15.917	0.0628	8.8633	0.1128	4.9219
13	1.886	0.5303	17.713	0.0565	9.3936	0.1065	5.3215
14	1.980	0.5051	19.599	0.0510	9.8987	0.1010	5.7133
15	2.079	0.4810	21.579	0.0464	10.3797	0.0964	6.0973
16	2.183	0.4581	23.658	0.0423	10.8378	0.0923	6.4736
17	2.292	0.4363	25.840	0.0387	11.2741	0.0887	6.8423
18	2.407	0.4155	28.132	0.0356	11.6896	0.0856	7.2034
19	2.527	0.3957	30.539	0.0328	12.0853	0.0828	7.5569
20	2.653	0.3769	33.066	0.0303	12.4622	0.0803	7.9030
21	2.786	0.3590	35.719	0.0280	12.8212	0.0780	8.2416
22	2.925	0.3419	38.505	0.0260	13.1630	0.0760	8.5730
23	3.072	0.3256	41.430	0.0241	13.4886	0.0741	8.8971
24	3.225	0.3101	44.502	0.0225	13.7987	0.0725	9.2140
25	3.386	0.2953	47.727	0.0210	14.0940	0.0710	9.5238
26	3.556	0.2813	51.113	0.0196	14.3752	0.0696	9.8266
27	3.733	0.2679	54.669	0.0183	14.6430	0.0683	10.1224
28	3.920	0.2551	58.403	0.0171	14.8981	0.0671	10.4114
29	4.116	0.2430	62.323	0.0161	15.1411	0.0661	10.6936
30	4.322	0.2314	66.439	0.0151	15.3725	0.0651	10.9691
31	4.538	0.2204	70.761	0.0141	15.5928	0.0641	11.2381
32	4.765	0.2099	75.299	0.0133	15.8027	0.0633	11.5005
33	5.003	0.1999	80.064	0.0125	16.0026	0.0625	11.7566
34	5.253	0.1904	85.067	0.0118	16.1929	0.0618	12.0063
35	5.516	0.1813	90.320	0.0111	16.3742	0.0611	12.2498
40	7.040	0.1421	120.800	0.0083	17.1591	0.0583	13.3775
45	8.985	0.1113	159.700	0.0063	17.7741	0.0563	14.3644
50	11.467	0.0872	209.348	0.0048	18.2559	0.0548	15.2233
55	14.636	0.0683	272.713	0.0037	18.6335	0.0537	15.9665
60	18.679	0.0535	353.584	0.0028	18.9293	0.0528	16.6062
65	23.840	0.0420	456.798	0.0022	19.1611	0.0522	17.1541
70	30.426	0.0329	588.529	0.0017	19.3427	0.0517	17.6212
75	38.833	0.0258	756.654	0.0013	19.4850	0.0513	18.0176
80	49.561	0.0202	971.229	0.0010	19.5965	0.0510	18.3526
85	63.254	0.0158	1245.087	0.0008	19.6838	0.0508	18.6346
90	80.730	0.0124	1594.607	0.0006	19.7523	0.0506	18.8712
95	103.035	0.0097	2040.694	0.0005	19.8059	0.0505	19.0689
100	131.501	0.0076	2610.025	0.0004	19.8479	0.0504	19.2337

APPENDIX C.10 6% Interest Factors

	Single Payment		Equal Payment Series				Uniform gradient-series factor
	Compound-amount factor	Present-worth factor	Compound-amount factor	Sinking-fund factor	Present-worth factor	Capital-recovery factor	
n	To find F Given P F/P i, n	To find P Given F P/F i, n	To find F Given A F/A i, n	To find A Given F A/F i, n	To find P Given A P/A i, n	To find A Given P A/P i, n	To find A Given G A/G i, n
1	1.060	0.9434	1.000	1.0000	0.9434	1.0600	0.0000
2	1.124	0.8900	2.060	0.4854	1.8334	0.5454	0.4854
3	1.191	0.8396	3.184	0.3141	2.6730	0.3741	0.9612
4	1.262	0.7921	4.375	0.2286	3.4651	0.2886	1.4272
5	1.338	0.7473	5.637	0.1774	4.2124	0.2374	1.8836
6	1.419	0.7050	6.975	0.1434	4.9173	0.2034	2.3304
7	1.504	0.6651	8.394	0.1191	5.5824	0.1791	2.7676
8	1.594	0.6274	9.897	0.1010	6.2098	0.1610	3.1952
9	1.689	0.5919	11.491	0.0870	6.8017	0.1470	3.6133
10	1.791	0.5584	13.181	0.0759	7.3601	0.1359	4.0220
11	1.898	0.5268	14.972	0.0668	7.8869	0.1268	4.4213
12	2.012	0.4970	16.870	0.0593	8.3839	0.1193	4.8113
13	2.133	0.4688	18.882	0.0530	8.8527	0.1130	5.1920
14	2.261	0.4423	21.015	0.0476	9.2950	0.1076	5.5635
15	2.397	0.4173	23.276	0.0430	9.7123	0.1030	5.9260
16	2.540	0.3937	25.673	0.0390	10.1059	0.0990	6.2794
17	2.693	0.3714	28.213	0.0355	10.4773	0.0955	6.6240
18	2.854	0.3504	30.906	0.0324	10.8276	0.0924	6.9597
19	3.026	0.3305	33.760	0.0296	11.1581	0.0896	7.2867
20	3.207	0.3118	36.786	0.0272	11.4699	0.0872	7.6052
21	3.400	0.2942	39.993	0.0250	11.7641	0.0850	7.9151
22	3.604	0.2775	43.392	0.0231	12.0416	0.0831	8.2166
23	3.820	0.2618	46.996	0.0213	12.3034	0.0813	8.5099
24	4.049	0.2470	50.816	0.0197	12.5504	0.0797	8.7951
25	4.292	0.2330	54.865	0.0182	12.7834	0.0782	9.0722
26	4.549	0.2198	59.156	0.0169	13.0032	0.0769	9.3415
27	4.822	0.2074	63.706	0.0157	13.2105	0.0757	9.6030
28	5.112	0.1956	68.528	0.0146	13.4062	0.0746	9.8568
29	5.418	0.1846	73.640	0.0136	13.5907	0.0736	10.1032
30	5.744	0.1741	79.058	0.0127	13.7648	0.0727	10.3422
31	6.088	0.1643	84.802	0.0118	13.9291	0.0718	10.5740
32	6.453	0.1550	90.890	0.0110	14.0841	0.0710	10.7988
33	6.841	0.1462	97.343	0.0103	14.2302	0.0703	11.0166
34	7.251	0.1379	104.184	0.0096	14.3682	0.0696	11.2276
35	7.686	0.1301	111.435	0.0090	14.4983	0.0690	11.4319
40	10.286	0.0972	154.762	0.0065	15.0463	0.0665	12.3590
45	13.765	0.0727	212.744	0.0047	15.4558	0.0647	13.1413
50	18.420	0.0543	290.336	0.0035	15.7619	0.0635	13.7964
55	24.650	0.0406	394.172	0.0025	15.9906	0.0625	14.3411
60	32.988	0.0303	533.128	0.0019	16.1614	0.0619	14.7910
65	44.145	0.0227	719.083	0.0014	16.2891	0.0614	15.1601
70	59.076	0.0169	967.932	0.0010	16.3846	0.0610	15.4614
75	79.057	0.0127	1300.949	0.0008	16.4559	0.0608	15.7058
80	105.796	0.0095	1746.600	0.0006	16.5091	0.0606	15.9033
85	141.579	0.0071	2342.982	0.0004	16.5490	0.0604	16.0620
90	189.465	0.0053	3141.075	0.0003	16.5787	0.0603	16.1891
95	253.546	0.0040	4209.104	0.0002	16.6009	0.0602	16.2905
100	339.302	0.0030	5638.368	0.0002	16.6176	0.0602	16.3711

APPENDIX C.11 7% Interest Factors

	Single Payment		Equal Payment Series				Uniform gradient-series factor
	Compound-amount factor	Present-worth factor	Compound-amount factor	Sinking-fund factor	Present-worth factor	Capital-recovery factor	
n	To find F Given P F/P i,n	To find P Given F P/F i,n	To find F Given A F/A i,n	To find A Given F A/F i,n	To find P Given A P/A i,n	To find A Given P A/P i,n	To find A Given G A/G i,n
1	1.070	0.9346	1.000	1.0000	0.9346	1.0700	0.0000
2	1.145	0.8734	2.070	0.4831	1.8080	0.5531	0.4831
3	1.225	0.8163	3.215	0.3111	2.6243	0.3811	0.9549
4	1.311	0.7629	4.440	0.2252	3.3872	0.2952	1.4155
5	1.403	0.7130	5.751	0.1739	4.1002	0.2439	1.8650
6	1.501	0.6664	7.163	0.1398	4.7665	0.2098	2.3032
7	1.606	0.6228	8.654	0.1156	5.3893	0.1856	2.7304
8	1.718	0.5820	10.260	0.0975	5.9713	0.1675	3.1466
9	1.838	0.5439	11.978	0.0835	6.5152	0.1535	3.5517
10	1.967	0.5084	13.816	0.0724	7.0236	0.1424	3.9461
11	2.105	0.4751	15.784	0.0634	7.4987	0.1334	4.3296
12	2.252	0.4440	17.888	0.0559	7.9427	0.1259	4.7025
13	2.410	0.4150	20.141	0.0497	8.3577	0.1197	5.0649
14	2.579	0.3878	22.550	0.0444	8.7455	0.1144	5.4167
15	2.759	0.3625	25.129	0.0398	9.1079	0.1098	5.7583
16	2.952	0.3387	27.888	0.0359	9.4467	0.1059	6.0897
17	3.159	0.3166	30.840	0.0324	9.7632	0.1024	6.4110
18	3.380	0.2959	33.999	0.0294	10.0591	0.0994	6.7225
19	3.617	0.2765	37.379	0.0268	10.3356	0.0968	7.0242
20	3.870	0.2584	40.996	0.0244	10.5940	0.0944	7.3163
21	4.141	0.2415	44.865	0.0223	10.8355	0.0923	7.5990
22	4.430	0.2257	49.006	0.0204	11.0613	0.0904	7.8725
23	4.741	0.2110	53.436	0.0187	11.2722	0.0887	8.1369
24	5.072	0.1972	58.177	0.0172	11.4693	o.0872	8.3923
25	5.427	0.1843	63.249	0.0158	11.6536	0.0858	8.6391
26	5.807	0.1722	68.676	0.0146	11.8258	0.0846	8.8773
27	6.214	0.1609	74.484	0.0134	11.9867	0.0834	9.1072
28	6.649	0.1504	80.698	0.0124	12.1371	0.0824	9.3290
29	7.114	0.1406	87.347	0.0115	12.2777	0.0815	9.5427
30	7.612	0.1314	94.461	0.0106	12.4091	0.0806	9.7487
31	8.145	0.1228	102.073	0.0098	12.5318	0.0798	9.9471
32	8.715	0.1148	110.218	0.0091	12.6466	0.0791	10.1381
33	9.325	0.1072	118.933	0.0084	12.7538	0.0784	10.3219
34	9.978	0.1002	128.259	0.0078	12.8540	0.0778	10.4987
35	10.677	0.0937	138.237	0.0072	12.9477	0.0772	10.6687
40	14.974	0.0668	199.635	0.0050	13.3317	0.0750	11.4234
45	21.002	0.0476	285.749	0.0035	13.6055	0.0735	12.0360
50	29.457	0.0340	406.529	0.0025	13.8008	0.0725	12.5287
55	41.315	0.0242	575.929	0.0017	13.9399	0.0717	12.9215
60	57.946	0.0173	813.520	0.0012	14.0392	0.0712	13.2321
65	81.273	0.0123	1146.755	0.0009	14.1099	0.0709	13.4760
70	113.989	0.0088	1614.134	0.0006	14.1604	0.0706	13.6662
75	159.876	0.0063	2269.657	0.0005	14.1964	0.0705	13.8137
80	224.234	0.0045	3189.063	0.0003	14.2220	0.0703	13.9274
85	314.500	0.0032	4478.576	0.0002	14.2403	0.0702	14.0146
90	441.103	0.0023	6287.185	0.0002	14.2533	0.0702	14.0812
95	618.670	0.0016	8823.854	0.0001	14.2626	0.0701	14.1319
100	867.716	0.0012	12381.662	0.0001	14.2693	0.0701	14.1703

APPENDIX C.12 8% Interest Factors

	Single Payment		Equal Payment Series				Uniform gradient-series factor
	Compound-amount factor	Present-worth factor	Compound-amount factor	Sinking-fund factor	Present-worth factor	Capital-recovery factor	
n	To find F Given P F/P i, n	To find P Given F P/F i, n	To find F Given A F/A i, n	To find A Given F A/F i, n	To find P Given A P/A i, n	To find A Given P A/P i, n	To find A Given G A/G i, n
1	1.080	0.9259	1.000	1.0000	0.9259	1.0800	0.0000
2	1.166	0.8573	2.080	0.4808	1.7833	0.5608	0.4808
3	1.260	0.7938	3.246	0.3080	2.5771	0.3880	0.9488
4	1.360	0.7350	4.506	0.2219	3.3121	0.3019	1.4040
5	1.469	0.6806	5.867	0.1705	3.9927	0.2505	1.8465
6	1.587	0.6302	7.336	0.1363	4.6229	0.2163	2.2764
7	1.714	0.5835	8.923	0.1121	5.2064	0.1921	2.6937
8	1.851	0.5403	10.637	0.0940	5.7466	0.1740	2.0985
9	1.999	0.5003	12.488	0.0801	6.2469	0.1601	3.4910
10	2.159	0.4632	14.487	0.0690	6.7101	0.1490	3.8713
11	2.332	0.4289	16.645	0.0601	7.1390	0.1401	4.2395
12	2.518	0.3971	18.977	0.0527	7.5361	0.1327	4.5958
13	2.720	0.3677	21.495	0.0465	7.9038	0.1265	4.9402
14	2.937	0.3405	24.215	0.0413	8.2442	0.1213	5.2731
15	3.172	0.3153	27.152	0.0368	8.5595	0.1168	5.5945
16	3.426	0.2919	30.324	0.0330	8.8514	0.1130	5.9046
17	3.700	0.2703	33.750	0.0296	9.1216	0.1096	6.2038
18	3.996	0.2503	37.450	0.0267	9.3719	0.1067	6.4920
19	4.316	0.2317	41.446	0.0241	9.6036	0.1041	6.7697
20	4.661	0.2146	45.762	0.0219	9.8182	0.1019	7.0370
21	5.034	0.1987	50.423	0.0198	10.0168	0.0998	7.2940
22	5.437	0.1840	55.457	0.0180	10.2008	0.0980	7.5412
23	5.871	0.1703	60.893	0.0164	10.3711	0.0964	7.7786
24	6.341	0.1577	66.765	0.0150	10.5288	0.0950	8.0066
25	6.848	0.1460	73.106	0.0137	10.6748	0.0937	8.2254
26	7.396	0.1352	79.954	0.0125	10.8100	0.0925	8.4352
27	7.988	0.1252	87.351	0.0115	10.9352	0.0915	8.6363
28	8.627	0.1159	95.339	0.0105	11.0511	0.0905	8.8289
29	9.317	0.1073	103.966	0.0096	11.1584	0.0896	9.0133
30	10.063	0.0994	113.283	0.0088	11.2578	0.0888	9.1897
31	10.868	0.0920	123.346	0.0081	11.3498	0.0881	9.3584
32	11.737	0.0852	134.214	0.0075	11.4350	0.0875	9.5197
33	12.676	0.0789	145.951	0.0069	11.5139	0.0869	9.6737
34	13.690	0.0731	158.627	0.0063	11.5869	0.0863	9.8208
35	14.785	0.0676	172.317	0.0058	11.6546	0.0858	9.9611
40	21.725	0.0460	259.057	0.0039	11.9246	0.0839	10.5699
45	31.920	0.0313	386.506	0.0026	12.1084	0.0826	11.0447
50	46.902	0.0213	573.770	0.0018	12.2335	0.0818	11.4107
55	68.914	0.0145	848.923	0.0012	12.3186	0.0812	11.6902
60	101.257	0.0099	1253.213	0.0008	12.3766	0.0808	11.9015
65	148.780	0.0067	1847.248	0.0006	12.4160	0.0806	12.0602
70	218.606	0.0046	2720.080	0.0004	12.4428	0.0804	12.1783
75	321.205	0.0031	4002.557	0.0003	12.4611	0.0803	12.2658
80	471.955	0.0021	5886.935	0.0002	12.4735	0.0802	12.3301
85	693.456	0.0015	8655.706	0.0001	12.4820	0.0801	12.3773
90	1018.915	0.0010	12723.939	0.0001	12.4877	0.0801	12.4116
95	1497.121	0.0007	18701.507	0.0001	12.4917	0.0801	12.4365
100	2199.761	0.0005	27484.516	0.0001	12.4943	0.0800	12.4545

APPENDIX C.13 9% Interest Factors

	Single Payment		Equal Payment Series				Uniform gradient-series factor
	Compound-amount factor	Present-worth factor	Compound-amount factor	Sinking-fund factor	Present-worth factor	Capital-recovery factor	
n	To find F Given P F/P i,n	To find P Given F P/F i,n	To find F Given A F/A i,n	To find A Given F A/F i,n	To find P Given A P/A i,n	To find A Given P A/P i,n	To find A Given G A/G i,n
1	1.090	0.9174	1.000	1.0000	0.9174	1.0900	0.0000
2	1.188	0.8417	2.090	0.4785	1.7591	0.5685	0.4785
3	1.295	0.7722	3.278	0.3051	2.5313	0.3951	0.9426
4	1.412	0.7084	4.573	0.2187	3.2397	0.3087	1.3925
5	1.539	0.6499	5.985	0.1671	3.8897	0.2571	1.8282
6	1.677	0.5963	7.523	0.1329	4.4859	0.2229	2.2498
7	1.828	0.5470	9.200	0.1087	5.0330	0.1987	2.6574
8	1.993	0.5019	11.028	0.0907	5.5348	0.1807	3.0512
9	2.172	0.4604	13.021	0.0768	5.9953	0.1668	3.4312
10	2.367	0.4224	15.193	0.0658	6.4177	0.1558	3.7978
11	2.580	0.3875	17.560	0.0570	6.8052	0.1470	4.1510
12	2.813	0.3555	20.141	0.0497	7.1607	0.1397	4.4910
13	3.066	0.3262	22.953	0.0436	7.4869	0.1336	4.8182
14	3.342	0.2993	26.019	0.0384	7.7862	0.1284	5.1326
15	3.642	0.2745	29.361	0.0341	8.0607	0.1241	5.4346
16	3.970	0.2519	33.003	0.0303	8.3126	0.1203	5.7245
17	4.328	0.2311	36.974	0.0271	8.5436	0.1171	6.0024
18	4.717	0.2120	41.301	0.0242	8.7556	0.1142	6.2687
19	5.142	0.1945	46.018	0.0217	8.9501	0.1117	6.5236
20	5.604	0.1784	51.160	0.0196	9.1286	0.1096	6.7675
21	6.109	0.1637	56.765	0.0176	9.2923	0.1076	7.0006
22	6.659	0.1502	62.873	0.0159	9.4424	0.1059	7.2232
23	7.258	0.1378	69.532	0.0144	9.5802	0.1044	7.4358
24	7.911	0.1264	76.790	0.0130	9.7066	0.1030	7.6384
25	8.623	0.1160	84.701	0.0118	9.8226	0.1018	7.8316
26	9.399	0.1064	93.324	0.0107	9.9290	0.1007	8.0156
27	10.245	0.0976	102.723	0.0097	10.0266	0.0997	8.1906
28	11.167	0.0896	112.968	0.0089	10.1161	0.0989	8.3572
29	12.172	0.0822	124.135	0.0081	10.1983	0.0981	8.5154
30	13.268	0.0754	136.308	0.0073	10.2737	0.0973	8.6657
31	14.462	0.0692	149.575	0.0067	10.3428	0.0967	8.8083
32	15.763	0.0634	164.037	0.0061	10.4063	0.0961	8.9436
33	17.182	0.0582	179.800	0.0056	10.4645	0.0956	9.0718
34	18.728	0.0534	196.982	0.0051	10.5178	0.0951	9.1933
35	20.414	0.0490	215.711	0.0046	10.5668	0.0946	9.3083
40	31.409	0.0318	337.882	0.0030	10.7574	0.0930	9.7957
45	48.327	0.0207	525.859	0.0019	10.8812	0.0919	10.1603
50	74.358	0.0135	815.084	0.0012	10.9617	0.0912	10.4295
55	114.408	0.0088	1260.092	0.0008	11.0140	0.0908	10.6261
60	176.031	0.0057	1944.792	0.0005	11.0480	0.0905	10.7683
65	270.846	0.0037	2998.288	0.0003	11.0701	0.0903	10.8702
70	416.730	0.0024	4619.223	0.0002	11.0845	0.0902	10.9427
75	641.191	0.0016	7113.232	0.0002	11.0938	0.0902	10.9940
80	986.552	0.0010	10950.574	0.0001	11.0999	0.0901	11.0299
85	1517.932	0.0007	16854.800	0.0001	11.1038	0.0901	11.0551
90	2335.527	0.0004	25939.184	0.0001	11.1064	0.0900	11.0726
95	3593.497	0.0003	39916.635	0.0000	11.1080	0.0900	11.0847
100	5529.041	0.0002	61422.675	0.0000	11.1091	0.0900	11.0930

APPENDIX C.14 10% Interest Factors

	Single Payment		Equal Payment Series				Uniform gradient-series factor
	Compound-amount factor	Present-worth factor	Compound-amount factor	Sinking-fund factor	Present-worth factor	Capital-recovery factor	
n	To find F Given P F/P i, n	To find P Given F P/F i, n	To find F Given A F/A i, n	To find A Given F A/F i, n	To find P Given A P/A i, n	To find A Given P A/P i, n	To find A Given G A/G i, n
1	1.100	0.9091	1.000	1.0000	0.9091	1.1000	0.0000
2	1.210	0.8265	2.100	0.4762	1.7355	0.5762	0.4762
3	1.331	0.7513	3.310	0.3021	2.4869	0.4021	0.9366
4	1.464	0.6830	4.641	0.2155	3.1699	0.3155	1.3812
5	1.611	0.6209	6.105	0.1638	3.7908	0.2638	1.8101
6	1.772	0.5645	7.716	0.1296	4.3553	0.2296	2.2236
7	1.949	0.5132	9.487	0.1054	4.8684	0.2054	2.6216
8	2.144	0.4665	11.436	0.0875	5.3349	0.1875	3.0045
9	2.358	0.4241	13.579	0.0737	5.7590	0.1737	3.3724
10	2.594	0.3856	15.937	0.0628	6.1446	0.1628	3.7255
11	2.853	0.3505	18.531	0.0540	6.4951	0.1540	4.0641
12	3.138	0.3186	21.384	0.0468	6.8137	0.1468	4.3884
13	3.452	0.2897	24.523	0.0408	7.1034	0.1408	4.6988
14	3.798	0.2633	27.975	0.0358	7.3667	0.1358	4.9955
15	4.177	0.2394	31.772	0.0315	7.6061	0.1315	5.2789
16	4.595	0.2176	35.950	0.0278	7.8237	0.1278	5.5493
17	5.054	0.1979	40.545	0.0247	8.0216	0.1247	5.8071
18	5.560	0.1799	45.599	0.0219	8.2014	0.1219	6.0526
19	6.116	0.1635	51.159	0.0196	8.3649	0.1196	6.2861
20	6.728	0.1487	57.275	0.0175	8.5136	0.1175	6.5081
21	7.400	0.1351	64.003	0.0156	8.6487	0.1156	6.7189
22	8.140	0.1229	71.403	0.0140	8.7716	0.1140	6.9189
23	8.954	0.1117	79.543	0.0126	8.8832	0.1126	7.1085
24	9.850	0.1015	88.497	0.0113	8.9848	0.1113	7.2881
25	10.835	0.0923	98.347	0.0102	9.0771	0.1102	7.4580
26	11.918	0.0839	109.182	0.0092	9.1610	0.1092	7.6187
27	13.110	0.0763	121.100	0.0083	9.2372	0.1083	7.7704
28	14.421	0.0694	134.210	0.0075	9.3066	0.1075	7.9137
29	15.863	0.0630	148.631	0.0067	9.3696	0.1067	8.0489
30	17.449	0.0573	164.494	0.0061	9.4269	0.1061	8.1762
31	19.194	0.0521	181.943	0.0055	9.4790	0.1055	8.2962
32	21.114	0.0474	201.138	0.0050	9.5264	0.1050	8.4091
33	23.225	0.0431	222.252	0.0045	9.5694	0.1045	8.5152
34	25.548	0.0392	245.477	0.0041	9.6086	0.1041	8.6149
35	28.102	0.0356	271.024	0.0037	9.6442	0.1037	8.7086
40	45.259	0.0221	442.593	0.0023	9.7791	0.1023	9.0962
45	72.890	0.0137	718.905	0.0014	9.8628	0.1014	9.3741
50	117.391	0.0085	1163.909	0.0009	9.9148	0.1009	9.5704
55	189.059	0.0053	1880.591	0.0005	9.9471	0.1005	9.7075
60	304.482	0.0033	3034.816	0.0003	9.9672	0.1003	9.8023
65	490.371	0.0020	4893.707	0.0002	9.9796	0.1002	9.8672
70	789.747	0.0013	7887.470	0.0001	9.9873	0.1001	9.9113
75	1271.895	0.0008	12708.954	0.0001	9.9921	0.1001	9.9410
80	2048.400	0.0005	20474.002	0.0001	9.9951	0.1001	9.9609
85	3298.969	0.0003	32979.690	0.0000	9.9970	0.1000	9.9742
90	5313.023	0.0002	53120.226	0.0000	9.9981	0.1000	9.9831
95	8556.676	0.0001	85556.760	0.0000	9.9988	0.1000	9.9889
100	13780.612	0.0001	137796.123	0.0000	9.9993	0.1000	9.9928

APPENDIX C.15 12% Interest Factors

	Single Payment		Equal Payment Series				Uniform gradient-series factor
	Compound-amount factor	Present-worth factor	Compound-amount factor	Sinking-fund factor	Present-worth factor	Capital-recovery factor	
n	To find F Given P F/P i,n	To find P Given F P/F i,n	To find F Given A F/A i,n	To find A Given F A/F i,n	To find P Given A P/A i,n	To find A Given P A/P i,n	To find A Given G A/G i,n
1	1.120	0.8929	1.000	1.0000	0.8929	1.1200	0.0000
2	1.254	0.7972	2.120	0.4717	1.6901	0.5917	0.4717
3	1.405	0.7118	3.374	0.2964	2.4018	0.4164	0.9246
4	1.574	0.6355	4.779	0.2092	3.0374	0.3292	1.3589
5	1.762	0.5674	6.353	0.1574	3.6048	0.2774	1.7746
6	1.974	0.5066	8.115	0.1232	4.1114	0.2432	2.1721
7	2.211	0.4524	10.089	0.0991	4.5638	0.2191	2.5515
8	2.476	0.4039	12.300	0.0813	4.9676	0.2013	2.9132
9	2.773	0.3606	14.776	0.0677	5.3283	0.1877	3.2574
10	3.106	0.3220	17.549	0.0570	5.6502	0.1770	3.5847
11	3.479	0.2875	20.655	0.0484	5.9377	0.1684	3.8953
12	3.896	0.2567	24.133	0.0414	6.1944	0.1614	4.1897
13	4.364	0.2292	28.029	0.0357	6.4236	0.1557	4.4683
14	4.887	0.2046	32.393	0.0309	6.6282	0.1509	4.7317
15	5.474	0.1827	37.280	0.0268	6.8109	0.1468	4.9803
16	6.130	0.1631	42.753	0.0234	6.9740	0.1434	5.2147
17	6.866	0.1457	48.884	0.0205	7.1196	0.1405	5.4353
18	7.690	0.1300	55.750	0.0179	7.2497	0.1379	5.6427
19	8.613	0.1161	63.440	0.0158	7.3658	0.1358	5.8375
20	9.646	0.1037	72.052	0.0139	7.4695	0.1339	6.0202
21	10.804	0.0926	81.699	0.0123	7.5620	0.1323	6.1913
22	12.100	0.0827	92.503	0.0108	7.6447	0.1308	6.3514
23	13.552	0.0738	104.603	0.0096	7.7184	0.1296	6.5010
24	15.179	0.0659	118.155	0.0085	7.7843	0.1285	6.6407
25	17.000	0.0588	133.334	0.0075	7.8431	0.1275	6.7708
26	19.040	0.0525	150.334	0.0067	7.8957	0.1267	6.8921
27	21.325	0.0469	169.374	0.0059	7.9426	0.1259	7.0049
28	23.884	0.0419	190.699	0.0053	7.9844	0.1253	7.1098
29	26.750	0.0374	214.583	0.0047	8.0218	0.1247	7.2071
30	29.960	0.0334	241.333	0.0042	8.0552	0.1242	7.2974
31	33.555	0.0298	271.293	0.0037	8.0850	0.1237	7.3811
32	37.582	0.0266	304.848	0.0033	8.1116	0.1233	7.4586
33	42.092	0.0238	342.429	0.0029	8.1354	0.1229	7.5303
34	47.143	0.0212	384.521	0.0026	8.1566	0.1226	7.5965
35	52.800	0.0189	431.664	0.0023	8.1755	0.1223	7.6577
40	93.051	0.0108	767.091	0.0013	8.2438	0.1213	7.8988
45	163.988	0.0061	1358.230	0.0007	8.2825	0.1207	8.0572
50	289.002	0.0035	2400.018	0.0004	8.3045	0.1204	8.1597

APPENDIX C.16 15% Interest Factors

	Single Payment		Equal Payment Series				Uniform gradient-series factor
	Compound-amount factor	Present-worth factor	Compound-amount factor	Sinking-fund factor	Present-worth factor	Capital-recovery factor	
n	To find F Given P F/P i, n	To find P Given F P/F i, n	To find F Given A F/A i, n	To find A Given F A/F i, n	To find P Given A P/A i, n	To find A Given P A/P i, n	To find A Given G A/G i, n
1	1.150	0.8696	1.000	1.0000	0.8696	1.1500	0.0000
2	1.323	0.7562	2.150	0.4651	1.6257	0.6151	0.4651
3	1.521	0.6575	3.473	0.2880	2.2832	0.4380	0.9071
4	1.749	0.5718	4.993	0.2003	2.8550	0.3503	1.3263
5	2.011	0.4972	6.742	0.1483	3.3522	0.2983	1.7228
6	2.313	0.4323	8.754	0.1142	3.7845	0.2642	2.0972
7	2.660	0.3759	11.067	0.0904	4.1604	0.2404	2.4499
8	3.059	0.3269	13.727	0.0729	4.4873	0.2229	2.7813
9	3.518	0.2843	16.786	0.0596	4.7716	0.2096	3.0922
10	4.046	0.2472	20.304	0.0493	5.0188	0.1993	3.3832
11	4.652	0.2150	24.349	0.0411	5.2337	0.1911	3.6550
12	5.350	0.1869	29.002	0.0345	5.4206	0.1845	3.9082
13	6.153	0.1625	34.352	0.0291	5.5832	0.1791	4.1438
14	7.076	0.1413	40.505	0.0247	5.7245	0.1747	4.3624
15	8.137	0.1229	47.580	0.0210	5.8474	0.1710	4.5650
16	9.358	0.1069	55.717	0.0180	5.9542	0.1680	4.7523
17	10.761	0.0929	65.075	0.0154	6.0472	0.1654	4.9251
18	12.375	0.0808	75.836	0.0132	6.1280	0.1632	5.0843
19	14.232	0.0703	88.212	0.0113	6.1982	0.1613	5.2307
20	16.367	0.0611	102.444	0.0098	6.2593	0.1598	5.3651
21	18.822	0.0531	118.810	0.0084	6.3125	0.1584	5.4883
22	21.645	0.0462	137.632	0.0073	6.3587	0.1573	5.6010
23	24.891	0.0402	159.276	0.0063	6.3988	0.1563	5.7040
24	28.625	0.0349	184.168	0.0054	6.4338	0.1554	5.7979
25	32.919	0.0304	212.793	0.0047	6.4642	0.1547	5.8834
26	37.857	0.0264	245.712	0.0041	6.4906	0.1541	5.9612
27	43.535	0.0230	283.569	0.0035	6.5135	0.1535	6.0319
28	50.066	0.0200	327.104	0.0031	6.5335	0.1531	6.0960
29	57.575	0.0174	377.170	0.0027	6.5509	0.1527	6.1541
30	66.212	0.0151	434.745	0.0023	6.5660	0.1523	6.2066
31	76.144	0.0131	500.957	0.0020	6.5791	0.1520	6.2541
32	87.565	0.0114	577.100	0.0017	6.5905	0.1517	6.2970
33	100.700	0.0099	664.666	0.0015	6.6005	0.1515	6.3357
34	115.805	0.0086	765.365	0.0013	6.6091	0.1513	6.3705
35	133.176	0.0075	881.170	0.0011	6.6166	0.1511	6.4019
40	267.864	0.0037	1779.090	0.0006	6.6418	0.1506	6.5168
45	538.769	0.0019	3585.128	0.0003	6.6543	0.1503	6.5830
50	1083.657	0.0009	7217.716	0.0002	6.6605	0.1501	6.6205

APPENDIX C.17   20% Interest Factors

	Single Payment		Equal Payment Series				Uniform gradient-series factor
	Compound-amount factor	Present-worth factor	Compound-amount factor	Sinking-fund factor	Present-worth factor	Capital-recovery factor	
n	To find *F* Given *P* F/P *i, n*	To find *P* Given *F* P/F *i, n*	To find *F* Given *A* F/A *i, n*	To find *A* Given *F* A/F *i, n*	To find *P* Given *A* P/A *i, n*	To find *A* Given *P* A/P *i, n*	To find *A* Given *G* A/G *i, n*
1	1.200	0.8333	1.000	1.0000	0.8333	1.2000	0.0000
2	1.440	0.6945	2.200	0.4546	1.5278	0.6546	0.4546
3	1.728	0.5787	3.640	0.2747	2.1065	0.4747	0.8791
4	2.074	0.4823	5.368	0.1863	2.5887	0.3863	1.2742
5	2.488	0.4019	7.442	0.1344	2.9906	0.3344	1.6405
6	2.986	0.3349	9.930	0.1007	3.3255	0.3007	1.9788
7	3.583	0.2791	12.916	0.0774	3.6046	0.2774	2.2902
8	4.300	0.2326	16.499	0.0606	3.8372	0.2606	2.5756
9	5.160	0.1938	20.799	0.0481	4.0310	0.2481	2.8364
10	6.192	0.1615	25.959	0.0385	4.1925	0.2385	3.0739
11	7.430	0.1346	32.150	0.0311	4.3271	0.2311	3.2893
12	8.916	0.1122	39.581	0.0253	4.4392	0.2253	3.4841
13	10.699	0.0935	48.497	0.0206	4.5327	0.2206	3.6597
14	12.839	0.0779	59.196	0.0169	4.6106	0.2169	3.8175
15	15.407	0.0649	72.035	0.0139	4.6755	0.2139	3.9589
16	18.488	0.0541	87.442	0.0114	4.7296	0.2114	4.0851
17	22.186	0.0451	105.931	0.0095	4.7746	0.2095	4.1976
18	26.623	0.0376	128.117	0.0078	4.8122	0.2078	4.2975
19	31.948	0.0313	154.740	0.0065	4.8435	0.2065	4.3861
20	38.338	0.0261	186.688	0.0054	4.8696	0.2054	4.4644
21	46.005	0.0217	225.026	0.0045	4.8913	0.2045	4.5334
22	55.206	0.0181	271.031	0.0037	4.9094	0.2037	4.5942
23	66.247	0.0151	326.237	0.0031	4.9245	0.2031	4.6475
24	79.497	0.0126	392.484	0.0026	4.9371	0.2026	4.6943
25	95.396	0.0105	471.981	0.0021	4.9476	0.2021	4.7352
26	114.475	0.0087	567.377	0.0018	4.9563	0.2018	4.7709
27	137.371	0.0073	681.853	0.0015	4.9636	0.2015	4.8020
28	164.845	0.0061	819.223	0.0012	4.9697	0.2012	4.8291
29	197.814	0.0051	984.068	0.0010	4.9747	0.2010	4.8527
30	237.376	0.0042	1181.882	0.0009	4.9789	0.2009	4.8731
31	284.852	0.0035	1419.258	0.0007	4.9825	0.2007	4.8908
32	341.822	0.0029	1704.109	0.0006	4.9854	0.2006	4.9061
33	410.186	0.0024	2045.931	0.0005	4.9878	0.2005	4.9194
34	492.224	0.0020	2456.118	0.0004	4.9899	0.2004	4.9308
35	590.668	0.0017	2948.341	0.0003	4.9915	0.2003	4.9407
40	1469.772	0.0007	7343.858	0.0002	4.9966	0.2001	4.9728
45	3657.262	0.0003	18281.310	0.0001	4.9986	0.2001	4.9877
50	9100.438	0.0001	45497.191	0.0000	4.9995	0.2000	4.9945

APPENDIX C.18 25% Interest Factors

	Single Payment		Equal Payment Series				Uniform gradient-series factor
	Compound-amount factor	Present-worth factor	Compound-amount factor	Sinking-fund factor	Present-worth factor	Capital-recovery factor	
n	To find F Given P F/P i, n	To find P Given F P/F i, n	To find F Given A F/A i, n	To find A Given F A/F i, n	To find P Given A P/A i, n	To find A Given P A/P i, n	To find A Given G A/G i, n
1	1.250	0.8000	1.000	1.0000	0.8000	1.2500	0.0000
2	1.563	0.6400	2.250	0.4445	1.4400	0.6945	0.4445
3	1.953	0.5120	3.813	0.2623	1.9520	0.5123	0.8525
4	2.441	0.4096	5.766	0.1735	2.3616	0.4235	1.2249
5	3.052	0.3277	8.207	0.1219	2.6893	0.3719	1.5631
6	3.815	0.2622	11.259	0.0888	2.9514	0.3388	1.8683
7	4.768	0.2097	15.073	0.0664	3.1611	0.3164	2.1424
8	5.960	0.1678	19.842	0.0504	3.3289	0.3004	2.3873
9	7.451	0.1342	25.802	0.0388	3.4631	0.2888	2.6048
10	9.313	0.1074	33.253	0.0301	3.5705	0.2801	2.7971
11	11.642	0.0859	42.566	0.0235	3.6564	0.2735	2.9663
12	14.552	0.0687	54.208	0.0185	3.7251	0.2685	3.1145
13	18.190	0.0550	68.760	0.0146	3.7801	0.2646	3.2438
14	22.737	0.0440	86.949	0.0115	3.8241	0.2615	3.3560
15	28.422	0.0352	109.687	0.0091	3.8593	0.2591	3.4530
16	35.527	0.0282	138.109	0.0073	3.8874	0.2573	3.5366
17	44.409	0.0225	173.636	0.0058	3.9099	0.2558	3.6084
18	55.511	0.0180	218.045	0.0046	3.9280	0.2546	3.6698
19	69.389	0.0144	273.556	0.0037	3.9424	0.2537	3.7222
20	86.736	0.0115	342.945	0.0029	3.9539	0.2529	3.7667
21	108.420	0.0092	429.681	0.0023	3.9631	0.2523	3.8045
22	135.525	0.0074	538.101	0.0019	3.9705	0.2519	3.8365
23	169.407	0.0059	673.626	0.0015	3.9764	0.2515	3.8634
24	211.758	0.0047	843.033	0.0012	3.9811	0.2512	3.8861
25	264.698	0.0038	1054.791	0.0010	3.9849	0.2510	3.9052
26	330.872	0.0030	1319.489	0.0008	3.9879	0.2508	3.9212
27	413.590	0.0024	1650.361	0.0006	3.9903	0.2506	3.9346
28	516.988	0.0019	2063.952	0.0005	3.9923	0.2505	3.9457
29	646.235	0.0016	2580.939	0.0004	3.9938	0.2504	3.9551
30	807.794	0.0012	3227.174	0.0003	3.9951	0.2503	3.9628
31	1009.742	0.0010	4034.968	0.0003	3.9960	0.2503	3.9693
32	1262.177	0.0008	5044.710	0.0002	3.9968	0.2502	3.9746
33	1577.722	0.0006	6306.887	0.0002	3.9975	0.2502	3.9791
34	1972.152	0.0005	7884.609	0.0001	3.9980	0.2501	3.9828
35	2465.190	0.0004	9856.761	0.0001	3.9984	0.2501	3.9858

Index